Factory Girl Literature

THE SEOUL-CALIFORNIA SERIES IN KOREAN STUDIES

The Seoul-California Series in Korean Studies is a collaboration between the University of California, Berkeley, and the Kyujanggak Institute for Korean Studies, Seoul National University. The series promotes the global dissemination of scholarship on Korea by publishing distinguished research on Korean history, society, art, and culture by scholars from across the world.

Factory Girl Literature

Sexuality, Violence, and Representation in Industrializing Korea

RUTH BARRACLOUGH

Global, Area, and International Archive
University of California Press
BERKELEY LOS ANGELES LONDON

The Global, Area, and International Archive (GAIA) is an initiative of the Institute of International Studies, University of California, Berkeley, in partnership with the University of California Press, the California Digital Library, and international research programs across the University of California system.

University of California Press, one of the most distinguished university presses in the United States, enriches lives around the world by advancing scholarship in the humanities, social sciences, and natural sciences. Its activities are supported by the UC Press Foundation and by philanthropic contributions from individuals and institutions. For more information, visit www.ucpress.edu.

University of California Press
Berkeley and Los Angeles, California

University of California Press, Ltd.
London, England

Library of Congress Cataloging-in-Publication Data

A catalog record for this book is available from the Library of Congress.

Manufactured in the United States of America

21 20 19 18 17 16 15 14 13 12
10 9 8 7 6 5 4 3 2 1

The paper used in this publication meets the minimum requirements of ANSI/NISO Z39.48–1992 (R 1997) (*Permanence of Paper*).

Contents

Preface and Acknowledgments

Over the ten years I have spent researching and writing this book, a curious problem has trailed it: the book's title conveyed quite a different meaning in Korean and in English. In Korea, my working title, *The Labor and Literature of Korean Factory Girls,* elicited groans from friends and peers, or worse, concerned silence. Factory girls had been studied, I was told. There was no hidden literature, it had all been ferreted out and published in the 1980s, and ample critical discussions of it had followed. My project appeared to loiter around the great street battles of the late 1980s, for which only those with a penchant for Leninist organizational minutiae felt any nostalgia or interest. So I changed the title to more accurately reflect the thread that guided me, choosing *Factory Girl Literature* instead. But this title jarred people too. The words *yŏgong* (female worker) and *munhak* (literature) simply do not belong together, I was told. I would need to provide some explanation, to soften the incongruity and abrasiveness created by the proximity of these two words. Literature and factory girls belong to two different worlds, separated by structures that invest each of them with histories that barely intersect. I would need to address this distance before creating and running with such a title.

By contrast, in Australia and the United States when I presented my work, people were kind enough to be charmed by the idea of a Korean factory girl literature. The great nineteenth-century industrial novels and the Korean manufacturing economy were equally distant here, and they had no power to unsettle an attempt at neat categorization.

Ever since the Welsh cultural critic Raymond Williams classified the mid-nineteenth-century novels of Dickens, Gaskell, Disraeli, and Kingsley as "industrial literature," the factory girl has taken her place as a key cultural figure in industrializing societies. She has traveled across nations, as well as genres, to inhabit an oeuvre that spans Europe, the United States, Australasia, Russia, and of course East Asia. Yet in Korea, rather than scrabbling to gain a prestige that only literature or the labor movement could confer, working-class women, when they took up their pens to write, confessed an uneasy relationship to writing and publishing. They seemed distrustful of a taxonomy that paid homage to them for their exotic distance from (and suitable longing for) culture and refinement. The relationship between factory girls and literature was always tense in these texts, as my friends had warned. The question, then, became how was it that some of these books, such as Kang Kyŏng-ae's *In'gan Munje* (The Human Predicament) and Shin Kyŏng-suk's *Oettan Bang* (The Solitary Room), classics of factory girl literature, entered the modern literary canon in Korea? Taking these problems to heart, this book examines the tensions between class, sex, and literature in Korea's industrialization experience—reading literature, not as the final arbiter of experience, but as one route by which new subjectivities might emerge.

Many people and institutions have aided in the writing of this book. My first acknowledgment must go to the Australian Student Christian Movement (ASCM) and the Korean Student Christian Federation (KSCF), who together in the 1980s ran an exchange program that sponsored university students from one country to undertake a political and religious "exposure tour" in the other. Reading though the ASCM archives recently, I learned that it was the feminist and democratic principles of the organization's executive committee that had them light upon a seventeen-year-old undergraduate from Queensland as a worthy recipient of the program scholarship in 1989. The KSCF had greater difficulty in sending female university students abroad, a reminder that in the tightly run military state of South Korea it was easier to leave university and go underground as a factory worker than to visit the sunburnt country. I would particularly like to thank John Ball, Russell Peterson, Marion Maddox, and my father for encouraging me to take that journey. In South Korea, Lee Eunju, Yun Youngmo, and Choi Jonga looked after me.

This book began as a PhD thesis undertaken at the Australian National University (ANU), and I would like to thank my supervisory committee for their support and encouragement and the inspiration of their own work. I thank Kenneth Wells for encouraging me right from the beginning and for giving generously of his time and expertise throughout the writing of the dissertation and beyond. I am grateful to Rick Kuhn for helping me think through the shift from dissertation to book and for his friendship when I returned to ANU to teach. I thank Aat Vervoorn for always being available when I needed to discuss my work and for his thoughtful insights. The three external examiners, Seung-kyung Kim, Hagen Koo, and Laurel Kendall, gave my thesis a rigorous, engaged and sympathetic appraisal; their ongoing support has meant a great deal to me.

While I was a PhD student at ANU, the following friends and colleagues provided valuable stimulation and encouragement: Kyoung-Hee Moon, Greg Evon, Andrei Lankov, Gi-Hyun Shin, Linda Bennett, Leonie Harcourt, David Hundt, Lewis Mayo, Kirill Nourzhanov, Linda Poskitt, and Vanessa Ward. In particular I thank Miriam Lang, who read my entire dissertation draft and offered extensive comments. Her encouragement and generous sharing of ideas was an integral part of my training.

In Korea I am indebted to many people and a number of organizations who helped me during my doctoral fieldwork: Professor Kim Dong-ch'un and the Labor Academy series at Songkonghwe University, the people of the Korean Federation of Public Sector Unions, Reverend Park Sang-jung and his late wife Reverend Lee Sun-ae, the people of the Seoul International Socialist Organization—especially Yong-ju—and the excellent Korean-language teachers at Yonsei, Ewha, and Seoul National University. For their hospitality and friendship I am grateful to Kang Song-won and Park Sang-i and their family, Kim Chi-min and her family, and Jillian Cheetham.

While I was writing, Chung Jin-ouk, Kyoung-hee Moon, and Minseon Lee all discussed factory girl literature with me, checked my translations, and saved me from overreading texts or from missing the point entirely. Their critique, encouragement, and friendship helped me greatly.

For its material support for my research, I am grateful to the Korea Foundation, which has transformed Korean studies in Australasia over the past fifteen years. While a Korea Foundation postdoctoral fellow at Sydney University, I was greatly aided by the collegiality of Elise Tipton, Ikuko Sorenson, Mami Iwashita, and Park Duksoo. During my brief sojourn in the History Department at the University of Minnesota, the friendship of Soojin and Alexs Pate, Jim and Rose Ryan, Ted Farmer, Ann Waltner, Barbara Welke, Jeffrey Pilcher, Hiromi Mizuno, and Chris Issett made our time there deeply rewarding. Returning to ANU as a faculty member in 2007, I have been very fortunate in my colleagues. Special mention must go to Bina D'Costa, Ronit Ricci, John Makeham, Kent Anderson, Judith Pabian, Hyaeweol Choi, Roald Maliangkay, Mark Gibeau, Duncan Campbell, and Gaik Cheng Khoo.

The incomparable Namhee Lee has been a wise and generous mentor since our first meeting in 2004. It was Namhee who introduced me to Nathan MacBrien at the University of California and took the time to nurture my work despite the many calls upon her time. Namhee, Laurel Kendall, Hwasook Nam, Elyssa Faison, Jiseung Roh, and Kim Won have all helped me think through the ideas that shape this book. Nathan MacBrien, my editor at the Institute for International Studies, University of California, Berkeley, and copyeditor Julie Van Pelt have done a beautiful job massaging my prose and getting me to clarify my ideas.

In Korea I am grateful to Kwŏn Yŏng-min, who has welcomed me into the extraordinarily stimulating and productive intellectual environment of the Korean Literature Department at Seoul National University. I have benefited immensely from his advice as well as from that of Professor Paik Nak-chung, who first urged me to read Shin Kyông-suk's *Oettan Bang*. I also thank my beloved friend Kim Hye-ran, who painstakingly taught me Korean in Seoul in 1992 and 1993 and whose friendship saw me through those early years in Seoul.

Although I began as a labor historian, my crossing into the field of Korean literature has been aided by a number of scholars whose advice and support it is my pleasure to acknowledge here: Choi Kyeong-Hee, Ji-Eun Lee, Yoon Sun Yang, Jin-Kyung Lee, and Ted Hughes. The two anonymous reviewers of this book for the University of California Press not only gave crucial suggestions on how to strengthen the manuscript but also outed themselves to follow up with ongoing advice and encouragement. I have poached their stimulating ideas wherever I could, and my deep thanks go to Paula Rabinowitz and Jin-Kyung Lee.

More than anyone else, Liam Dee has helped me think through and write this book. He has read and discussed every part of it with me, tried to improve the clarity of my arguments, and given generously of his own ideas. In material terms he has suffered obstacles to his academic career in order to be with our daughter and me. My daughter, Una, also deserves credit for distracting me with the toil and pleasure of looking after her.

My parents, Reverend Ray and Dorothy Barraclough, deserve acknowledgment for nurturing the interests that drive this book. Discussions, now decades long, we have shared about literature, radical politics, and gender relations have helped shape this work. For this, and much else besides, I dedicate this book to them.

An earlier draft of chapter 2 appeared as "Tales of Seduction: Factory Girls in Korean Proletarian Literature," *positions: east asia cultures critique* 14, no. 2 (Fall 2006). Chapter 4 appeared in an earlier form in *Gender and Labour in Korea and Japan: Sexing Class*, edited by Ruth Barraclough and Elyssa Faison and published by Routledge in 2009. I am grateful to the publishers of these works for allowing me to reprint this material here.

Introduction

Sexuality, Violence, Literature

The steam whistle screamed its six o'clock tidings and factory girls in their soiled whites came shivering out into the evening air. So many women—fourteen and fifteen year olds with their hair tied back in kerchiefs, young tired mothers in their late twenties, all eager to chatter and look about them as though released into a glorious freedom. Since seven in the morning they had wrestled with the cotton spinning machines and now their portion of slavery was over. Sloughing off their cramped factory body and stepping into their freeborn selves, they departed for the evening. Mokpo, famous for its coastal breezes and dramatic scenery, became at this time of the day infused with a beautiful glow as the cotton dust from the factories turned the setting sun a deep red. They saw it with satisfaction, and as the sea breeze brushed their tired and sweaty faces it seemed to make them beautiful once more.

PAK HWA-SŎNG, "Chusok Chŏnya" (The Night before Chusok)

So opens Pak Hwa-sŏng's 1926 debut story, which won her, at age twenty-two, the attention of Korea's literary world. Published in *Chosŏn Mundan* (Korean Literary World), the preeminent literary magazine of Korea's new modern writers, "The Night before Chusok" is notable for many reasons. In dexterous and forceful language, Pak establishes the factory districts as a crucial site for modern literature in Korea. She relates the violent sexual dynamics of the factories with a clarity and urgency that gives veracity to the scenes she depicts. The main character of the story, Yŏngshin, is introduced returning home with an injured arm, the outcome of her passionate defense of a young worker being sexually harassed by the line manager. Yŏngshin herself is one of the "tired young mothers" who works at the factory, a widow who supports her young son and mother-in-law. Yŏngshin is established as a lone woman, whose husband's death has exposed her to the vagaries of labor practices in the colonial economy. Yet she herself is a target of sexual violence also, and the story seems intent on focusing on her isolation and vulnerability. Right from its earliest incarnation, factory girl literature seemed to seek out the lone woman, the country girl in Seoul, surrounded not by friends and comrades but by predators who enjoyed almost complete power over her actions. The question for us is, why was this so?

This book examines the process by which factory girls became cultural figures of immense political significance in modern Korean literature and its industrializing society. Their importance to the economy in key periods of industrialization can hardly be overstated. In the 1920s and 1930s they dominated the textile industries, and from the early 1930s until the end of World War II they made up over 30 percent of the industrial workforce, laboring in textile mills, silk-reeling factories, rubber-shoe production, rice mills, and tobacco factories. In the second great wave of industrialization, enacted under military rule in the 1960s and 1970s, laboring women were the conerstone of Korea's resurgent manufacturing economy. By the 1980s over a million women were commuting to factories in cities and towns all over South Korea, where they produced clothes, toys, candy, and electrical gadgets for the global market. Just as laboring women generated the manufacturing economy of Korea, they also helped create a factory girl discourse that was used to evaluate the capitalist experience.

As factories created the aggrievements, which in turn inspired literature, the factory girl emerged as a cultural figure in her own right. Part of larger moves for the enfranchisement of the working class, the figure of the factory girl combined in her person a disturbing and powerful mix of the political and the sexual. The rendering of the factory girl as a sentimental figure in literature speaks both to a paternalist stream in the labor movement and to a wider aesthetic fixation on feminine suffering. This politics of sentiment gave working-class women a role—equal parts prominent and debased—within the labor movement of the 1920s and 1930s and later in the 1970s and 1980s. Rather than focusing on portraits of factory girls authorized by the literary world, *Factory Girl Literature* analyzes the costs of this discursive tradition for working-class women themselves. As an examination of the cultural moment when working-class women began to write, this book explains why female loneliness and alienation, even in the midst of the mass labor movement of the 1980s and 1990s, is the dominant trope of their work.

By asking how literary representation came to substitute for the political representation of working-class women, this book demonstrates the interrelationship between politics, sex, emancipatory movements, and their narratives. Sentimental factory girl literature worked to control women in the 1920s just as these women were entering the new social space of the factory. When factory girls appeared in literature as objects of pity, rather than enhancing their cultural authority, it underlined their reliance on the patronage of the compassionate reader. By tying the illiterate lower-class woman to the literary figure who spoke of all that was wrong with colonial capitalism, the factory girl achieved her rhetorical power via economic and cultural structures that curbed and silenced her. This book connects this literary representation of factory girls with wider discourses on female labor, working-class politics, and what feminist labor historian Anna Clark has termed "the trauma of industrialisation."[1] Above all this book is about what it means to be a working-class woman in relation to literary culture. If Virginia Woolf in her famous aphorism was correct that the moment when the middle-class woman began to write shaped the making of the modern world,[2] what can the writing of working-class women tell us about modernity, capitalism, and ourselves?

WHO IS THE FACTORY GIRL?

Korea first experienced industrialization as a colony of Japan (1910–1945), and by the late 1920s young Korean women were entering the light-manufacturing industries in significant numbers. The term used to describe these women in Korean, *yŏgong* (여공), or "female worker," is very different from the antiquated English term I use throughout this book, "factory girl." How do I substantiate this translation? *Yŏgong* is a term that at first glance appears to be completely functional. It is composed of the characters *yŏ* (女 female) and *gong* (工 work). *Gong* may, strictly speaking, apply to any kind of labor, but the connotation has been built up through years of association with work in factories. Labor historian Yi Ok-ji traces the first appearance of the term *yŏgong* to newspaper accounts of two strikes in 1919, one at a rice mill in Pusan and the other in a rice-polishing factory in Inch'ŏn.[3] Thus, from the very beginning, employers and, above them the architects of industrial development, had difficulty fixing the meaning of this term. The pithiness of *yŏgong* meant that it could be used to announce job vacancies in the factories and to broadcast the opening of new mills. At the same time, radical journalists, socialist writers, and labor activists were borrowing it to alert the public to the abuses that went on in the female-dominated factories. Almost from the moment of coining, the meaning of *yŏgong* fluctuated and expanded. Fiction writers' turn to examine the interior world of *yŏgong* coincided with factory women writing in to newspapers and magazines. Together, over the 1920s and early 1930s, these two groups filled out this new identity in letters and stories redolent with pathos and political energy. By the early 1930s, *yŏgong* was a cultural marker as much as an economic category, a figure that resonated with archetypal force in a period of immense socioeconomic change.

For the translator the question thus becomes one of both accuracy and evocation. Writing between languages, I use the English term "factory girl" to make a deliberate connection between one body of industrial literature and another. The social-problem fiction of the 1840s and 1850s that Raymond Williams famously classified as "industrial literature" is an important point of comparison for this study. Williams argued that even though the industrial novels were driven by a fear of revolution, they were always interested in representing, and sympathizing with, working-class characters.[4] In colonial Korea, however, which is our staring point for examining the first appearance of *yogong,* the political terrain was both harsher and more receptive to arguments for the radical reorganization of society. After the nation was lost to the colonial takeover by Japan in 1910, socialist, communist, and anarchist ideas and movements gained prominence. They took inspiration and support from political developments elsewhere in East Asia and the success of the Bolshevik revolution in Russia. Moves to industrialize the Korean peninsula to make it a productive colony for Japan, and the creation of a new urban class of day laborers and contract factory workers, coincided with the introduction into the colony of a new discourse of proletarian power.

The sympathy and suspicion that the image of the female proletarian triggered in Korea found expression in the ambiguity with which she was represented. She was a sexual object though not a feminine one, an erratic figure who threatened to scorn the impulses of pity and disapproval with which she was regarded. Upon her were fixed the hopes and dreams of the two rival forces of twentieth-century Korea: the communists and the capitalists. But the clichés peddled by these two rival camps do not concern us here. This is not a book about the impoverished discourse on factory girls. Nor is it an exercise in "history from below," a way of analyzing that relies on the permanence of the very hierarchies it professes to critique.[5] It does not find political validation in the status of marginalization, nor does it accept literary prominence as a substitute for poor wages, scrappy meals, and disenfranchisement. Instead this book inquires into the deepest ambitions of factory girls and rereads these texts to discover not only what they desired but where they found pleasure and authority. It attempts to look beyond the homage intrinsic to so much factory girl representation and searches instead for moments of clarity and autonomy by which we can educate ourselves anew on life in an industrial revolution.

The figure of the factory girl in modern Korean literature is riven by contradictions. Burdened with the capacity to convey all the horrors of the factory system, factory girls came to be tainted by association with those abuses. They were condemned as unfeminine while also exposed to extreme sexual harassment in the factories and on the streets of industrializing Seoul. Signifying both gender ambiguity and sexual availability, working-class women wrote of being caught in a social world that appeared determined to exclude and exploit them. Yet the factory girl, with all her ambiguities, was also hoisted to a remarkable prominence by labor and socialist movements that endowed her with political significance. This book is possible only because of the rich archival material that those movements spawned, generating a discourse on factory girls that goes into minute detail about the economic and social realities of working-class life.

This book argues that factory girls became a crucial marker in modern Korean literature for the violence of the rapid industrialization experience. They were depicted as susceptible to the labor market's deceit and vulnerable to the factory system's violence in ways that established their political consequence. As the degredation of factory girls became part of their literary equipage, chroniclers of factory girl literature snagged on how to depict a working-class feminine subjectivity. Specifically, how do you depict factory girls as sexual subjects in an environment overwhelmed by the violence of primitive capital accumulation? And, working within social realism, how do you create characters that do more than ritually enact their own degredation?

Faced with these questions, class conflict came to be managed as sexual violence in factory girl literature. In the rape scenes of 1930s proletarian literature, and the extreme sexual harrassment found in the autobiographies of the 1970s, sexual violence and workplace deception became the marker by which readers perceived the heroine's class. Over time the representation of proletarian women and of sexual violence became mutually constitutive. Sexual violence not only became the dramatic trope of the factory girl, it worked to solicit readers' interest in her.

Faced with this conundrum, how does one conduct a rereading and reevaluation of factory girl literature? In translating and discussing these works, how is one to analyze both the message and the context of these stories? At its heart this book argues that the violence and the tedium, the seductions and the aggrievements of industrialization united to produce new female working-class subjectivities. Factory girl literature, even with its limitations, enabled and showcased the emergence of new working-class selves in industrializing Korea. The process by which this happened, and its consequences for working-class women, is the subject of this book.

THE SCHOLARLY CONTEXT

My search through the archives of Korean labor history and modern literature has been directed by the questions posed in two books by Jacques Rancière. In *The Nights of Labor,* Rancière demonstrated a new kind of labor history devoted to uncovering the exhilarating meeting of workers and intellectuals in associations driven by a particular context of class curiosity, over a narrower search bent on locating a sealed-off "working-class culture." Rancière's arguments in *The Philosopher and His Poor* made me question to what degree the poignancy of the factory girl as literary heroine was due to her exclusion from the act of writing literature. Rancière's theory of "exclusion by homage" has provided an effective tool to reread what had once seemed paradoxical in the history of modern Korean literature: suppression of the writing of factory girls alongside their power as literary figures.

A number of books have explored ideas of literature, labor, class, and sexuality in ways that have been very fruitful for this study. In her book on factory girls in Victorian literature, *Hidden Hands: Working-Class Women and Victorian Social-Problem Fiction,* Patricia Johnson examines the tensions between gender and class that factory girls embodied in industrializing England. She shows how working-class women were a problem for industrial novelists like Dickens, Gaskell, and Disraeli, because in Victorian ideology laboring women are neither truly feminine nor authentic workers. I argue that these gender norms played themselves out in Korea, whereby femininity and labor came to be defined as antithetical, and I examine the consequences of this lack of a sense of stable femininity in factory girl literature.

Paula Rabinowitz's *Labor and Desire: Women's Revolutionary Fiction in Depression America* is exemplary for showing how one might write a gendered history of literary radicalism. By detailing how gender is rendered in radical literature and how female sexuality was written into working-class narratives, Rabinowitz defines the potential of radical narratives for depicting a female working-class subjectivity otherwise lost. Her book first showed me how one might bring together in one coherent whole the study of radical movements, texts, and female sexuality. Nancy Armstrong's book, *How Novels Think,* on the emergent individual in the first modern novels, provides crucial insights on how novels make particular kinds of selfhood possible. When writing the final chapter on the creation of factory girl interiority in canonical literature, I used as my guide Armstrong's analysis of how desire that threatens the social order is disciplined in the act of creating the modern individual in fiction.

On the subject of working-class characters in literature, Amy Schrager Lang's *The Syntax of Class* provides a method for how one might read class in a wider social context dominated by a fear of class conflict. This question is a particularly fruitful one for factory girl literature penned in the 1970s and 1980s in the anticommunist state of South Korea. The suggestion that class conflict is managed in literature through particular tropes—in America, of racial and gender difference; in Korea, through a trope of sexual danger—provides a compelling template by which to read class in Korea's industrial literature. Finally, Morag Shiach's *Modernism, Labour and Selfhood* greatly expands the context for thinking through the development of the modern category of labor in cultural forms. Though concerned with the British experience, Shiach's work raises important questions about the iconography of working-class movements whose visual representations of working women vacillate between showing them as either asexual and unthreatening or sexually vulnerable. In Korea the problem of how to represent the factory girl was dominated by concerns over authenticity and respectability and the demands of social realism. Rather than seeking to resolve these dilemmas, this book occupies the space created by these conflicts and reads, through the fates of their characters, the outcome of these debates.

In South Korea, labor history has been an important area for feminist historians since the 1970s and 1980s. When Yi Sŏng-hŭi wrote that "the women's movement began again in the labor movement of the 1970s,"[6] she was acknowledging the tangled history of women's organizations in South Korea and the important role that women's labor disputes in the 1970s played in challenging the elite nature and priorities of Korea's women's organizations up until that time.[7] The women's labor movement in the 1970s ignited a radical feminism that was deeply felt in intellectual circles also. The impact of the working-class women's movement that campaigned for rights specific to their gender—protesting the firing of married women workers, of women who had become pregnant, and defending their women union leaders—was echoed in the early feminists' interest in and commitment to labor history.

Lee Hyo-chae's analysis of the circumstances of young women in factories in the colonial period, "Ilcheha-ŭi Yŏsŏng Nodong Munje" (The Situation of Women Workers in the Colonial Period), opened up the field when it was published in 1978; it introduced the lives and struggles of colonial-era factory girls to a society where many of the same brutal practices applied in the factory districts of Seoul. When Chŏng Hyŏn-baek published the article "Yŏsŏng Nodongja-ŭi Ŭisik kwa Nodong Segye: Nodongja Suki

Punsŏkŭl Chunsimŭro" (Women Workers' Consciousness and the World of Work: Analyzing Workers' Writings), in the first volume of the feminist journal *Yŏsŏng* (Women) in 1985, her work stood alongside a bourgeoning labor movement as she examined society through the writings of working-class women. Labor history scholars were responding both to the growing strength of labor unions as the 1980s progressed and to the sense of urgency that the besieged, often illegal unions were experiencing. Commonly, labor historians in the 1980s and early 1990s dedicated their books and articles to the democratic union movement *(minju nojo undong),* thereby tying the aspirations of contemporary workers and unionists to the struggles of the colonial-era labor movement.[8] These South Korean historians and literary critics, who had themselves lived through late industrialization, would look back to the earlier industrial experience to reconstruct a broken genealogy of the struggles of working-class people.

In Korea, factory girl literature has had two distinct heydays, the first in proletarian literary circles in the 1920s and 1930s and the second as part of a dissident labor movement in 1970s and 1980s South Korea.[9] In both periods, factory girl literature was part of larger moves for the enfranchisement of the working class. In the late 1920s and early 1930s, Korea's proletarian literature movement addressed the concerns of the new class of industrial laborers who were leaving behind the old agricultural economy to fill the ports and cities of colonial Korea. In the 1980s a new *nodong munhak,* or "labor literature" movement, emerged just as the growing self-confidence of the working class in South Korea was making itself felt. This book connects these two distinct periods in Korean history to examine the changing portrayal of factory girls in literature and to chart the rise of working-class women authors. As a genealogy of working-class women's literature, this study is also an attempt to recover some of the lost voices of the Korean industrialization experience. Only fragments of published writing by working-class women survive from the colonial period, because literacy was so rare in the factories and many women were afraid of retribution or were simply "silenced by ordinary labor."[10] Yet the concerns of working-class women in the 1920s and 1930s were taken up by authors of the proletarian literature movement, whose heroines are a valuable source on the lives and aspirations of factory girls and of those who sought to emancipate them through literature. By the 1970s and 1980s, literature by female factory workers had become more plentiful, both because of increasing literacy and a flourishing dissident literary market that in the 1980s thrived on the popularity of labor literature. In chapters 3 and 4, I concentrate on the three most famous works of autobiographical labor literature by factory girls, who critiqued their society in intimate tales of the class divide.

After a decade of indifference and neglect on the subject of factory girls in Korea, in 2006 Kim Won published his magisterial *Yŏgong 1970: Kŭ-nyŏdul-ŭi Pan Yŏksa* (1970 Factory Girls: A Counterhistory). By critically evaluating the factory girl discourses of the 1960s, 1970s, and 1980s that were circulated by dissident intellectuals and company representatives, and bringing to the fore the voice of working-class women themselves, Kim Won offered a critical perspective on the female labor movement of the 1970s. His book shares with Seung-kyung Kim's *Class Struggle or Family Struggle* a critical evaluation of the women's labor movement of the 1970s. Long perceived as a tragically heroic episode in Korean labor history, the factory girl strikes of the late 1970s occupy an important place in histories of worker-state relations in the 1970s, at the end of the Park Chung Hee regime. By asking what the women won for themselves out of these "heroic" and often futile and brutal campaigns, both writers question the grand narrative of labor history and reposition factory girls at the center of their own experience. This approach has influenced my own research and has steered me toward close textual analysis of factory girl writings. Though they have been celebrated for the "selfless" and "inspirational" political risks they took in taking on capital and the state, factory girls' own conflicted selfhood has been overlooked.

Factory Girl Literature shares with Namhee Lee's book, *The Making of Minjung: Democracy and the Politics of Representation in South Korea,* a concern with the politics of representation in South Korea. Lee explores the struggle of South Korean intellectuals to "recover subjectivity" from the crises of national division, military rule, late industrialization, and a neocolonial relationship with the United States. This journey led to the extraordinary alliance of intellectuals and workers over the 1970s and 1980s and the creation of a new genre of "labor literature." In a book packed with insight, Lee shows what was at stake for intellectuals in this process. *Factory Girl Literature,* by contrast, turns its attention to the factory girl writers who worked to bring about a "common culture" in South Korea in these years. But it also connects the counterculture of the 1970s and 1980s with its important precursor, the "red decade" of the colonial period.

THE STRUCTURE OF THE BOOK

Beginning with the first mention of factory girls *(yŏgong)* in newspapers in 1919, in chapter 1 I examine radical journalism, letters to newspapers, articles, opinion pieces, anonymous poetry, and strike notices to show how the female-dominated factories of the 1920s were understood as sites of sexual violence, either as part of the disciplining of the female workforce or as evidence of the lack of control exercised upon labor relations. This was borne out on the factory floor by the wage and fine system, which compelled female operatives to comply with the culture of sexual harassment in the factories. Piecing together fragmentary records penned by factory girls that depict the first decades of colonial industrialization, I show how working-class women were some of the first and most ambivalent interpreters of the seductions of capitalist relations.

Socially engaged literature and working-class radicalism were two features of the 1920s and early 1930s, both a product of "colonial modernity" in Korea and a critique of it. Turning to works from the proletarian literature avant-garde of the 1920s and 1930s, chapter 2 examines the representation of working-class women in the proletarian literature movement (1925–35), where factory girls were depicted as sexual victims of capitalism, seduced or raped by factory overseers. I show how these tales were a response to the rampant sexual violence in the female-dominated factories, without necessarily being a repudiation of said violence. These seduction tales reproduced in fictional form the turmoil and crisis of a rapid industrialization carried out by a coercive, imperialist state. But they also depicted a particular kind of radicalism in which suffering heroines became the erotic object of socialism, a trope that would have repercussions for factory girl literature in future decades.

The retreat of the factory girl from literature after the 1930s until well into the 1970s was part of her departure from political life also. In chapter 3 I chart that decline and the emergence of a new working-class female figure in the culture of the labor movement in the 1970s. At the same time that the new genre of the working-class autobiography was revitalizing South Korea's cold war culture wars, a resurgent labor politics was using the suffering of factory girls as a key moral argument against the authoritarian developmental state. Yet the cost of taking on the role of "suffering heroines" was immense. I examine a flash point in the burgeoning labor movement of the 1970s—the nude demonstration at Tongil in 1976—to illustrate how destructive this role could be.

Chapter 4 examines the educational structures of South Korea in the 1970s to explain the exhausting and ingenious process by which autodidact working-class women began to enter the literary world. In particular I focus on the relationship between manual labor and cultural labor as a central component in the growing alliance between workers and students that in the 1980s worked to overthrow the long reign of the military presidents. To shed light on this coalition I conduct a close reading of the work of the three most acclaimed working-class women authors, Sŏk Chŏng-nam, Chang Nam-su, and Song Hyo-sun. I connect their self-representation to the work of authors who influenced them: Thomas Hardy and Heinrich Heine, among others. And I relate their critique of the intimate effect of the class divide in stories of class romance, of the consolations of literature, and of the separation of families.

In the 1990s, as South Korea's prosperity and democratization appeared to make redundant the language of labor radicalism, in literature a new movement of retrospection was coming to terms with the costs of rapid industrialization and the revolutionary movements that had opposed it. Defining this retrospective moment was the writer Shin Kyŏng-suk, whose critically acclaimed and best-selling novel, *Oettan Bang* (The Solitary Room), found a writer at the height of her powers recollecting her adolescent years spent working in a factory. Narrating the gap between proletarian author and factory girl heroine, Shin's book has been seen as the inheritor of Korea's factory girl literature tradition. Chapter 5 conducts a rereading of *The Solitary Room* and evaluates its turn toward interiority, girl love, and suicide in the factories.

All in all this project aims to enlarge our understanding of how working-class women interacted with the labor and literary movements that sought to represent them. It critically analyzes the discourses surrounding factory girls to highlight how the political power attributed to working-class women was based on the romanticization of their suffering. It shows that this could occur because the factory girl entered modern Korean literature in the 1920s at a time when the writing of lower-class women was suppressed. And it goes on to explore what working-class women themselves made of this literary tradition when they took up their pens from the 1970s to write. Through an analysis of literature in its social context, this project excavates the gap between working-class women as social actors and factory girls as cultural figures. I argue that the novels of the early proletarian literature experiment of the 1920s and 1930s, the factory girl autobiographies of the 1970s and 1980s, and the work of acclaimed novelist Shin Kyŏng-suk in the 1990s, are archetypal accounts of an intense historical experience that make up a distinct archive I call factory girl literature.

Factory girl literature is best understood at the intersection of labor politics and literary movements. This literature is much more than a straightforward narrative of the rapid industrialization experience. Rather it brings to light how cultural prominence can be mistaken for political authority and exposes the ambivalence inherent in projects of cultural enfranchisement that reinforce the authority of literature even as they open up new worlds of experience. As it turned out, factory girls and the fictions created about them tell us a great deal about life in an industrial revolution. They show how painful it was to be a heroine of the labor movement and an underling in one's own family, to be alone in the capital and sexually curious, to be hungry for literature but without the space or time to read. The questions that tormented them about their own capacity as writers are in this book turned into important questions about the limits of representation in the overlapping worlds these factory girls inhabited.

1. The Invention of the Factory Girl

> Such an awful difference between hearing about women's lives and work [and actually seeing it]. . . . They say working in the cotton spinning factories or on the telephone exchange is a decent job for lower-class women, and moreover when I visited the girls I received the impression that their work was not too arduous. But when I went to the rice mills and the rubber factories I saw mothers carrying a baby on their back as they toiled. What is this wretched thing called work?
>
> ANONYMOUS "LADY JOURNALIST," in *Shin Kajong* (New Family)[1]

> The overseer stood beside the fresh factory girls who had just arrived, laughing and chatting to them. And giving one girl's plump buttocks a resounding slap he cried, "Work hard! That's how you make a bonus!"
>
> KANG KYŎNG-AE, *In'gan Munje* (The Human Predicament)

From 1919, when the category of *yŏgong* (여공), or "factory girl," first appeared in newspapers, journalists, essayists, novelists, and poets attempted to represent this new modern type. Questions abounded: Who were the new factory girls? And what did they want? When the first factories were established, from 1910 through the 1920s, were factory girls simply availing themselves of the opportunities opened to them by a new capitalist labor market? Or were they pressed by poverty into work contracts that continued the old bonded labor system under a veneer of modernity? And when these women appeared in the pages of the new daily newspapers and magazines, whom could readers trust to represent them? Was it the "lady journalists" *(puin kija)*, self-confessed tourists to the working class who brought back horrified accounts of the rubber factories such as the one that opens this chapter? Or the anonymous poetry and opinion pieces, signed by "a factory girl," that pleaded with readers for a radical charity that might rescue them from the fortress factories? While the factory girl haunted proletarian literature and radical journalism during this early period of industrialization, she remained tantalizingly inaccessible to her sympathetic readers. This is because the figure of the factory girl was being constructed as an economic, social, and sexed figure at a time when illiteracy was standard in the factories. The question then becomes, how do we evaluate the representation of factory girls at a time when the writing of working-class women was suppressed?

The political appeal made by factory girl literature rested on the productive relationship between reportage and fiction, as news stories of strikes, lockouts, and country girls tricked into brothels in the mid- to late 1920s made their way into short stories and novels. Yet the factory girl as literary subject was created at a time when laboring women and girls had very little access to literary production. This word "access" separates those with the means, however meager, to give themselves time and study to write, from those who had to devote all of their energies to scraping out a living for themselves and any dependents. The scraps of writing that we do have are from those who lost a husband or a father and as a result fell from one class down to another, into an employment society of considerable complexity. In some instances new occupations appeared before any clear status had been affixed to these women, thus the circumstance in the chapter epigraph of a lady journalist in search of "decent" jobs for lower-middle-class and lower-class women.

Laboring women's own published writings were an important part, but still only a portion, of the composite "factory girl." However, their literary interventions conveyed crucial clues to the writers and activists who sought to represent them more systematically as the 1920s progressed. Focusing on the figure of the factory girl that emerged in the reportage and literary ephemera of this period illuminates how people navigated early capitalism in Korea. Did they attach their own private aspirations and dreams to the new economic and social spaces opening up for young women in the colony, or did they apprehend the hardship and degradation? Radical journalism took up this question, as reporters roamed factories and field strikes trying to gauge the contours of industrial life. Their reports were shaped by the context of colonial capitalism, an enterprise in empire building fiercely contested by its subjects in Korea. And because industrialization and socialism arrived in the Korean colony around the same time, the figure of the factory girl became discursively powerful out of all proportion to her numbers. Connected to the antagonistic enterprises of capitalist development and socialism, but with the pathos of their sex, factory girls in literature eventually emerged to do the political work that others in this economy could not do. But the crucial point here is that the factory girl became significant in modern Korean literature not despite her illiteracy but because of it. Her muted voice, her capacity to express in visceral suffering what others could only convey theoretically, infinitely expanded the range of her expressive power. This chapter examines the invention of the category of factory girl both in the female labor market and in the popular press, showing how the creation of a modern female proletarian subjectivity was as much a literary act as an economic one.

CAPITALIST EXPANSION

The first systematic program of industrialization in Korea began during the country's years as a colony of Japan, and by the late 1920s young Korean women were entering factories in significant numbers. As of 1931 the three largest industries in the Korean colony—textiles, food processing, and chemicals—employed large numbers of women in their plants and factories: 79 percent of the workforce in textiles was female, and women occupied 30 percent of jobs in both food processing and chemicals.[2] While in 1931 factory workers employed in the new, "modern" enterprises constituted only a small percentage of the total population in what was largely an agricultural economy, by the end of Japan's colonization in 1945 Korea possessed the industrial infrastructure from which South Korea's own version of rapid industrialization would be launched in the 1960s.

When Korea was opened to the international capitalist market of the late 1870s, Japanese entrepreneurs were some of the first to develop business interests in Korea, interests that would complement Japan's own import and export needs. The result was a rudimentary market economy both dependent on and companion to Japan's; by the time Japan formally colonized Korea on August 22, 1910, feudal and agricultural Korea was already taking in goods manufactured in Japan and exporting rice to feed the new factory workers of the industrializing colonizer.[3]

The period of Japan's colonial occupation of Korea (1910–1945) saw the most extensive development of capitalist industrialization until the 1960s. When large-scale industrialization commenced in Korea after 1919, it followed several decades of extreme rural poverty.[4] The agricultural economy that had served the Chosŏn dynasty (1392–1910) so long was collapsing under the weight of a corrupt and rapacious tax system. In these years many peasant farmers gave up their farms, unable to meet the the triple burden of a land tax, corvée labor, and grain tax that local officials tried to extort from them. Another cause of rural impoverishment was the Japanese cadastral survey completed in 1918, which overlooked the customary rights of tenant farmers and led to many smallholders losing their

..a farmers become vagrants, while others migrated with ..o Manchuria or Siberia. In 1910 the Chosŏn dynasty col- ..a domestic impoverishment and national uncertainty.

..ne Korean countryside, swelling impoverishment had created an ..rsupply of farm laborers and unemployed, who drifted between the rural and urban labor makets looking for temporary work. The circumstance of thousands of people seeking work in factories, mines, and seaports through recruiters in the 1920s was a direct consequence of the poverty of the agricultural economy. From the late Chosŏn dynasty period, many peasant farmers were landless and worked on fields they leased from others, or they sold their labor to bigger farms for wages. Those "rural proletarians" who came to seek work in the cities were leaving the insecurity of farm labor in the provinces, where job queues were long and the wages fluctuated seasonally.[6]

This shift in the composition of a rural proletarian class, and the migration to the cities, would eventually be cemented in the rigours of the war economy of the 1940s.[7] Industrialization was aided by the construction of a national railway in Korea by the Japanese military railway bureau, part of a transport network so expansive that by 1945 one could travel by train "from Pusan to Paris."[8] While Korea remained a predominantly agricultural economy throughout the colonial period, capitalist growth in the cities was swift. In 1911 there had been approximately 250 factories in Korea. By 1920 there were over 2,000,[9] and in 1943 over half a million people were working in factories across Korea.[10] But despite the large output of many of the industries, in the 1920s most factories operated on a very small scale. In 1925 the average number of employees in a factory was just eighteen. And a 1927 inquiry into the working conditions of factories in western Seoul reported that "they may be called factories but, with the exception of one or two really large establishments, most of what passes for a factory is really a shack. None of them seem to possess a lavatory."[11]

Women as well as men were making the journey to industrial towns and port cities in the 1920s to find work as servants, factory girls, waitresses or to train as nurses. The variety of jobs open to men was greater than what was available to females—men sought work as day laborers, miners, porters, stevedores, construction workers, servants, or factory hands. The prime nonagricultural industries of the 1920s and 1930s that employed lower-class women were the rice mills, cotton-spinning and silk-weaving factories, and rubber factories. Female participation in factory work would peak in 1934 at 34 percent of all waged workers.[12] Some industries specialized in the employment of female factory hands. In 1931,

71 percent of employees in the cotton-spinning industry were female, and among them were many girls—child workers—between the ages of twelve and fifteen.[13]

In many sectors women made up the majority of the workforce throughout the 1920s and 1930s. In addition to their high participation in the cotton-spinning industry, women constituted 86 percent of the workforce in the silk-weaving factories and 67 percent in the rubber industries.[14] Labor historian Yi Jŏng-ok reports that in addition to employing a predominantly female workforce, the silk-reeling factories also employed a high number of child workers under the age of sixteen.[15] In 1921, 66 percent of silk-factory employees were child workers, at a time when about 20 percent of the employees at cotton-spinning and -weaving factories were children under sixteen.[16]

The colonial period was a time of industrial growth that reshaped employment patterns. Korea's famous "farm to factory"[17] proletarianization began in the 1920s and 1930s with the rural exodus to industrial centers inside Korea and to the wider imperial labor market in Japan and Manchuria. But people did not always uproot willingly. The economic historian Carter Eckert has described the practice of poor farmers sending their daughters to work in factories as "an act of desperation."[18] Others have recorded that when recruiters came to the villages, families hid their daughter's shoes and clothes and locked them away to stop them from going out to work in the factories.[19] As recruiters fanned out into the countryside to find young, strong females to work at the larger establishments, such as the Japanese-run Chosen Textile Company or the Korean Kyŏngsŏng Spinning and Weaving Company, reports were already reaching the provinces through the newly established daily newspapers of abusive and dangerous work practices.

In colonial Korea there was no minimum wage, no suffrage, and no legislation to regulate the factory system. Japan's own Factory Law, promulgated in 1916 after "thirty-odd years of discussion," brought the following benefits to workers: "[It] prevented women and children under the age of fifteen from working more than twelve hours a day, prevented them from working between the hours of 10pm and 4am, and allowed for at least two days' leave per month. The minimum age for workers was set at twelve years, except in light industry, where the minimum age of workers was ten."[20] These provisions did not apply to Korea. When the Factory Law was revised in the 1920s, and night shift for women workers was abolished in Japan, a number of large Japanese cotton-spinning companies shifted their operations to Korea, where they were unencumbered by regulations.[21]

The wage system was tiered in colonial Korea. Japanese male workers who worked in Korea were paid the highest and occupied the uppermost positions on the factory floor as foremen and managers; Japanese women were paid on average half that of Japanese men, an amount equivalent to Korean male workers; Korean female workers were paid roughly half the wages paid to Korean men (and Japanese female workers); and Korean child workers were paid on average half the wages of Korean adult female workers.[22] Although wages fluctuated throughout the 1920s and 1930s, and fell sharply after the 1929 stock market crash, average wages stayed within this wage structure. The sector that reportedly paid the lowest wages to female workers was the silk-reeling industry, which in 1931 paid an average of 10 sen a day to its lowest-paid workers. The silk-industry average in this year was 41 sen a day, and women working in textiles earned an average of 60 sen a day.[23]

At this time, workers in Japanese and Korean factories could be punished for many vaguely defined "infringements"; overseers exerted considerable power over workers using the threat of a fine, of overtime, of physical punishment, or of firing. Historian Patricia Tsurumi details a list of transgressions that were punishable by fines in a cotton-spinning factory in Japan. The list included such vague imputations as "improper behaviour," "bad conduct stemming from laziness," and "obscene behaviour."[24] Carter Eckert describes a similar bonus system in the Kyŏngsŏng Spinning and Weaving Company in Korea that allowed workers to be fined for minor infractions.[25] The fines were imposed arbitrarily, with no possibility for appeal, and featured in the grievances listed in strike notices at major industrial disputes throughout these years.

The bonus and penalty system that augmented or docked wages was at the heart of the factory's discipline regimes. As labor historian Janice Kim has shown, penalty fines *(polgum)* incensed workers because they moved the cost of faulty products onto the workers when the factory's poor tools and machinery were to blame.[26] But bonuses were the chief measure of keeping female workers close and tractable. The bonus, a euphemism that obscures how essential this pay supplement was in making up a living wage, tied workers to the factory ethos. It was calculated based on attitude and compliance as well as production targets, and it was usually administered at the discretion of the supervisors. It was a policy set up to tie the women to a culture of complicity with long hours, injurious conditions, sexual harassment, and in some cases snitching.

Carter Eckert goes into considerable detail to estimate the average monthly wage of a young woman working at the Kyŏngsŏng Spinning and

Weaving Company in 1930, and the figure he reaches is "beggarly low" at roughly 15 yen per month.[27] Many girls entered factories because their families could not support them, and could not afford a dowry, but the amounts that factory girls earned kept them firmly within the working poor. Far from giving them independence, or entrée into the new world of consumer capitalism, the wage structure reminded factory girls that their best chance of survival lay in marriage.

Indeed the remuneration system at big factories like the Kyŏngsŏng Spinning and Weaving Company shows an intimate knowledge of the narrow needs of young working women, in particular their need to save money for themselves and often their families too, while supporting themselves away from home. They paid for their lodging and their meals at the factory, as well as their clothes and amusements, out of their small salaries. The wage system in Korea that discriminated against females and children was loosely based on the notion of a "family wage," whereby a working-class male was paid at a higher rate to enable him to keep his family at home.[28] The resultant wage discrimination against single women and lone children drawn into the factory economy underlines how flawed this policy was.

THE FEMALE LABOR MARKET

When young women entered the colony's rice mills, textile factories, match factories and rubber and tobacco factories in the 1920s and early 1930s, they usually had been recruited from farms and rural villages, where the cheapest labor was to be found. Factory girls were thus some of the first "moderns" of colonial Korea, as they left village communities and families to join their fate to the vagaries of the capitalist market. In the labor-intensive industries of the 1920s and 1930s, cheap, unskilled female labor found its niche. While servant girls and farm hands exemplified the labor patterns of the old economy, lower-class females in cities encountered employment opportunities in the new fields of the factory and the modern brothel.[29]

To facilitate the lucrative trade in female labor, a new breed of recruiters *(mojibin)* operated at the big provincial railway stations in Korea, buying and selling young women for factory work. The flourishing traffic in women was a direct consequence of the poverty in the Korean countryside, which launched thousands of girls and women into the search for city and port-town jobs. "If you went to Pusan Station there were always swarms of people selling factory girls," a contemporary described the scene. "There were also cases of these people selling factory girls privately, and if the company's price was right sometimes they would sell twenty or thirty girls at a time."[30] Lulled by the promise of monthly earnings and protection in factory dormitories, young women and girls sometimes traveled far to remote textile factories where their dependence on the company would be absolute. But the female labor market was a multilayered, maze-like system. Newspapers carried tales of girls lured by recruiters and then sold to faraway taverns or brothels, sometimes as far away as Japan.[31]

The prostitution industry was not only an alternative employment avenue for factory girls, it also influenced wider social anxieties about lower-class women going out to work as isolated, unprotected females. And, indeed, in the recruitment and sale of labor, the sex industry and the manufacturing industry in the early colonial period had much in common. Both used a bonded-labor system that sold the individual's labor power for a set number of years. Moreover, to be looking for work in the factories was easily conflated with being sexually available. Not only recruiters but journalists played a role in establishing these connections in readers' minds. The road to the factory was strewn with pitfalls, and enigmatic dangers awaited inside the factory itself. Sexual violence, or rather, the license for violence inherent in a bonded-labor economy, became the marker by which newspaper readers comprehended the class of the women who shared their cities.

In newspaper articles, magazines, anonymous poetry, letters to editors, fiction published as diaries, and other pieces of radical journalism and reportage, and eventually in short stories and novels from the proletarian literature movement, an overwhelming theme presented itself whenever factory girls appeared: sexual violence in the factories. Whether as an arbitrary measure of punishment (being publicly stripped for running away), or enmeshed in the wage system (where bonuses were purchased by entering into sexual relationships), or within a wider culture that saw laboring young women as huntable creatures, the threat of sexual violence infused coverage of working-class women. Yet the various genres had different things to say about the culture of sexual harassment in the factories. Two genres are closest to the source: strike notices and letters.

Strike notices, with their list of grievances and demands, appear to be the most reliable and transparent, the least mediated of all of the genres that reported on factory girls. In some notices a clear allegation was made. At the Chikbo Factory in Mokpo in 1926, 140 workers stopped work to make a single demand: that sexual violence by factory foremen should cease.[32] Again in 1926, at a rice mill, female workers went on strike demanding an end to arbitrary assaults on factory girls, requesting that one particular manager be replaced. At another rice mill in Inch'ŏn, the workers went on strike, "enraged at the Japanese manager's beating of pretty, young factory girls." At the Sunota Clothing Company, in a strike in 1926, among the grievances was "the sexual assault of factory girls by the Japanese supervisor."[33]

From 1926, cases of sexual harassment and assault in the factories first began to appear as stories in their own right in the *Tonga Ilbo* (East Asia Daily), a national daily newspaper established in 1920. Also in 1926, Pak Hwa-sŏng published her factory girl tale that delved into the physical and sexual intimidation experienced by factory hands. From this time on, control of the image of factories began to pass out of the hands of factory promoters and managers and into the hands of critics. Also in 1926 the leading progressive journal of the day, *Kaebyŏk* (Creation), published this poem anonymously, titled "Factory Girl":

> Although spring is here winter lingers
> in the sleeping streets of tired souls.
> The first factory siren breaks the dawn
> sears the blue-black sky
> and howls in the ears of three hundred thousand.
> I race to the factory breathing painfully
> unable to set out the mourning altar for my husband
> who died, crushed in a harness.
> In that evil cave
> I stoop for twelve hours and do not once turn my head
> to see that loathsome foreman making eyes at me.
> But in this foul world I must take whatever comes
> Oh my ancestors! My husband!
> Why did you leave me behind?[34]

Grief mingles with the resentments of poverty here. Without the breadwinner's wage, usually double her own, a lone woman would find it difficult to bypass the bonuses that were a vital part of her wages. But the bonus system tied women to complicity with a factory culture of long hours, injurious conditions, and sexual harassment. This poem also shows how women used the language available to them to critique their circumstances. Deeply resenting her own powerlessness in an economy that stripped her of self-governance, the author tells of her longing for her husband.

The part that male coworkers played in the opposing the harassment of working-class women is not always clear. I found only one case of male and female workers together protesting the sexual harassment of women workers. But there are other examples of male and female workers acting together, such as the two strikes in Inch'ŏn in 1931 that took place in rice mills, where male and female workers went out on strike together to demand, among other things, that women workers receive wages equal to men.[35] And Kang Kyŏng-ae reminds us, in her novel of the 1930s, that male and female coworkers were frequently drawn to each other. Yet abuses existed here also. In 1930, in P'yŏngyang, six hundred workers at the Sanship Silk Reeling Company went on strike to demand a ten-hour day, edible food, and that male coworkers should cease their practical jokes *(namkong-ŭi nongdamŭl kŭmhara).*[36] First-hand descriptions of sexual coercion in the factories, when they appeared, were damning. In November 1929, the *Tonga Ilbo* published the following letter in its regular column on working women, "Chigŏp Puini Doegi Kajji" (Becoming Career Women). The author gave her name as Yi Sŏng-ryong:

> I was three and my brother was seven when we lost our father and our mother went to work in a mill. I graduated from Normal School at fifteen and my brother, who had also graduated from Normal School and was working as a factory hand at a tailors, was twenty when he fell ill and died, leaving me to follow the same route and become a factory girl in a tobacco company. That was the spring I turned seventeen. My wages were 10 sen a day, but during the three weeks of apprenticeship I earned 6 sen a day, so in a month I earned about 30 sen. But if you wanted to flirt with the supervisor or the foreman you could earn double that a day, while if you rubbed someone up the wrong way you'd be swallowing abuse the whole time and suffering all manner of indignities. Those girls only earned 20 sen.[37]

As must have happened to many others, the death of a father spelled ruin for this family, who fell a class and entered factories to support themselves. In Yi Sŏng-ryong's account, we glimpse the management style in a tobacco factory where female workers are employed both to work hard and to appeal to their male supervisors. She goes on:

> To me the factory was like a lion's lair, and going to work was as hateful as being a cow going off to the slaughterhouse. Every day the male workers tried to waylay us in their heavy hunger for sex. But that wasn't the only thing. Every day when we clocked off, the guards frisked us one by one as if we were criminals.
>
> Readers, do not be shocked. I was a seventeen-year-old girl and that brutish supervisor would run his hands over my breasts until he reached the lower part of my body. How mortified I was. A seventeen-year-old girl forced to submit her body to this brutish handling. For 30 sen in wages I held back tears of blood and passed three long years of endless seasons.

Yi Sŏng-ryong writes openly of sexual assault in her workplace and accuses factory supervisors as well as working-class male colleagues of harassment and intimidation. At a time when there was no word for sexual harassment,[38] Yi Sŏng-ryong spells out in language that no reader can mistake the treatment that working-class women were subjected to in the factories. Most important of all, she reveals how factory girls were drawn into complying with the culture of abuse in the factories. When one's monthly wages depended on, among other things, a subjective criterion of "good behavior" and obedience, who could repel the overseer's overtures?

> At the beginning of the autumn I turned nineteen, I was overjoyed to hear that "if you go to X textile factory in Pusan the apprenticeship period is three months and discounting the food expenses you get 15 yen, and after three months you can earn an average of 50 yen," and for some reason I became awfully happy. . . . Readers, do not be surprised. Once I arrived in Pusan I couldn't see for tears. The food they said was for eating was foreign rice[39] and tofu stew, and the workday was twelve hours long. The work was two shifts, day and night. And even in the heat of summer, with the temperature hitting 90 degrees, the doors to the factory had to be kept shut. The reason they gave for this was that the air coming in would snap the threads. It was normal for the supervisor to be some kind of a bastard. If you couldn't show a comely face then, like in the tobacco factory, you got a terrible time. We never saw the 30 sen they promised to give us for our apprenticeship. . . . After plunging ten fingers into boiling water [day after day], my hands lost their beauty and many times I stroked them weeping. Readers, whether we received 30 sen or 1 yen, would this pittance make much difference to one family? It would be nothing. . . . Readers, take this one factory girl's complaint to your hearts, and believing that through your efforts we may one day listen to a happier tale, I lay down my pen.

Virginia Woolf's comment, upon reading Florence Nightingale's thinly veiled autobiography, that it was "not like writing, more like screaming" seems appropriate to this slice of autobiography also.[40] Two months after this epistle was printed, the 2,207 workers of the Pusan All-Korea Cotton Spinning Factory walked out on strike to protest a plan to reduce their wages. Their grievances included the poor quality of factory food, the penalty system, discrimination against Korean workers, and the treatment of women workers.[41] Protests over the mistreatment of women workers appear again and again in strike notices from this period. One of the strike demands at the Pusan factory—that women workers should not be prevented from leaving the factory grounds[42]—indicated the degree of control that management tried to exert over women workers. While male factory hands had some freedom of movement, factory girls were issued a ticket upon which their entries and departures were recorded, an indication that managers felt they could control the women completely.

In time the representation of factory girls and the representation of sexual violence began to encroach upon one another. Candid acknowledgment of a culture of sexual intimidation in the factories went hand in hand with vagueness over how such a system actually worked. Yet comprehending sexual violence in the factories is the key to understanding why factory girls were represented in literature from this period as sexual victims of capitalism. The opacity of the bonus and fine system, the lack of regulations on recruitment and job contracts, the distortions in the female labor market, and the extreme vulnerability of young, sturdy women in grueling jobs all served to mystify some of the leading principles of the factory system. And if the factory's punishment system was vague and arbitrary, sexual harassment was also messy and idiosyncratic. If the factory girl, despite all the effort and curiosity that went into depicting her, remained an elusive figure in journalism and literature, surely that too is due to the ambiguous politics of complicity and punishment that ordered capitalist relations in the factories.

Clearly, raising the problem of sexual violence in the factories was one thing, understanding the function of gender violence in the factory system quite another. Euphemisms stood in for details of mistreatment both in strike notices and in newspaper accounts of factory disputes. And while proletarian literature in the early 1930s developed a mission to expose the power relations masked by euphemisms and general bourgeois dissembling, it largely failed to deal with the sexuality of young waged women. Sexual violence in proletarian literature served as a clear warning, but whether to the factory girl, to her sympathetic readers, or to her tormentors was hard to tell.

In sum, whether Yi Sŏng-ryong was a "real" factory girl or a figment of an editor's imagination, her appeal was voiced in the language of serfdom, and her plea was for a charitable justice. Here we see the dominant power of precapitalist relations, as people caught up in the trauma of industrialization harked back to their "feudal, idyllic patriarchal ties" for deliverance. Yi Sŏng-ryong was asking to be rescued from the trap of poverty, foul working conditions, and sexual harassment that she intimated was the life of factory girls. In her search for a rescuer, a patron, she called upon the readers of newspapers for aid, not her fellow workers. Nevertheless, Yi

Sŏng-ryong's important innovation was in the mode of her appeal—she used the modern mass media to address a whole nation of readers and in so doing was one of the first to claim the public sphere for the concerns of factory girls.

INDUSTRIALISM AND REPRESENTATION

First and foremost, reporters, writers, and editors needed to establish a *tone* for writing about female industrial labor. Would it be scientific or sentimental? Horrified or sympathetic? Charitable or strictly objective? Coverage of the first major industrial action taken by female workers offers some clues. In July 1923, women workers at the Kyŏngsŏng Rubber Factory in Seoul elected an all-female trade union and walked off the job to protest falling wages and to demand the firing of an abusive overseer.[43] Their strike was the first major industrial action taken by female laborers in the colonial period. It made waves throughout the colony as newly established trade unions, labor clubs, youth groups, and radical organizations rallied to the female strikers' support, collecting solidarity money, sending telegrams of support, and organizing lecture tours to aid their cause.[44] The strike in the field outside the factory was chronicled in the pages of the *Tonga Ilbo*:

> Around 150 workers wipe the sweat pouring off them as they [stand] under the fierce midday sun and are drenched by passing storms. Before they have time to dry their dripping hemp skirts and *choksam* jackets, the evening dew has fallen and these 150 women [prepare to] spend the night under the boughs of acacia trees. . . . The firm stance of the factory's management shows no sign of yielding as they refuse to allow the women even water. Those women between the ages of twenty and thirty withstand these terrible circumstances thanks to their youth, but the older women in their forties and fifties, and the children under fifteen years of age, weep and try to persevere. Finally, when they become weak they lie down in the field where the night breeze stirs the grass.[45]

This slice of colonial journalism reminds us that strikes create subjectivities, but not straightaway. The journalist seems to be feeling his or her way, searching for the appropriate register with which to depict the strikers. In the florid, didactic language that characterized newspapers and magazines of the time (dubbed "social missionaries" by translator and literary historian Marshall Pihl), the writer enters eagerly into the suffering of the striking women. Yet while the article has drama and pathos, it is a sketch oddly devoid of power. Developing instead a theme of what I call "radical charity," early coverage of factory girl strikes reads as remarkably subjective reportage, as the boundaries that sympathy creates were being worked out. This eagerness to enter into the experience of laborers is a marked feature of journalism and literature of this period. The Kyŏngsŏng Rubber Factory strike was the beginning of a long string of factory girl strikes that pitted these workers against their employers and the police and brought them publicity and a new distinction in Korea's burgeoning capitalist society. Whether the distinction belonged to them or to the chroniclers who endeavored to pluck them free from mute labor, we will now examine.

This period in Korean history is significant to historians of working-class women not only because of the importance of female labor to the industrialization project, nor for the many incidences of factory girl strikes, but also because these were the years when writers and journalists turned a new, critical attention to their own society, and newspapers and magazines began to include accounts of the lower classes. This important cultural shift took place in the years of Governor-General Saito Makoto's "cultural period" and its aftermath. Named the "cultural period" for the remarkable flowering of national cultural production in literature, music, film, and art and the important political role that culture assumed in this time—both as a repository of "Koreaness" and for articulating political positions when other more conventional political forums were closed—it lasted until the early 1930s, when the Japanese army started expanding into China and colonial rule moved onto a war footing.[46]

Following the March First Independence demonstrations in 1919, and the retaliation of the colonial police, Japan altered its tactics in the colony, replacing a repressive police state with a more conciliatory, and effective, governing style.[47] In fact, the atmosphere in Korea in the 1920s and early 1930s in many ways mirrored that in Japan itself, where, according to historians Robert Scalapino and Chong-Sik Lee, "Japanese authorities tended to make a distinction between the right to engage in radical political action and the right to speak or write in a radical style."[48] As radical journalism in Korea thrived, and as a proletarian literature movement emerged—stimulated by political and literary developments in Japan, Russia, and other parts of the world—the literary world began to engage with the concerns of a new class of waged workers, including factory girls.

Colonial Korea's two major dailies, the *Tonga Ilbo* and the *Chosŏn Ilbo* (Chosŏn Daily), were attentive to the growing labor movement throughout the 1920s and early 1930s. Established in 1920, when the colonial government relaxed its restrictions on Korean-language newspapers and magazines,[49] these two papers played a vital role in providing an arena for public debate around some of the major issues of the time. Due to restrictions in the political realm—the absence of a parliament, of political parties or political representation, of any formal, public way of engaging with or influencing the colonial government—newspaper writing took on a significant role in these years for the communication of ideas and as a medium for trying to mobilize segments of the population.

Historian Michael Robinson has shown how eclectic the two papers were in the years of Governor-General Saito's cultural policy. Established by businessmen of "moderate" political views,[50] the two papers expressed sentiments that covered the political spectrum, from editorials that reproduced the owners' bourgeois nationalism to serialized proletarian literature. The papers published the outcomes of wildcat strikes, recorded the confederation meetings of mushrooming trade unions, and listed the demands of striking workers in major and minor disputes up and down the country.[51]

The history of the formation of these newspapers is also a history of the formation of a modern reading public in colonial Korea. Readers in the 1920s and early 1930s were treated to a constellation of different publications. Kim Kyŏng-il quotes the fictional serial "Diary of a Young Socialist," published in the magazine *Chosŏn Chikwang* (Light of Chosŏn) in 1927, in which the author communicates excitement about ideas sweeping the colony, while at the same time satirizing the "radical chic" of young socialist intellectuals:

> The day after tomorrow is the evening lecture. I'm the speaker, and I still haven't given a single thought to what I'm going to talk about, what'll I do? And the subject is awfully grand, "The Liberation Movement and the Economic Status of Women." I have to talk on that and I haven't a clue, what will I do? What can I say to the audience? They come expecting a lot of learning from these lectures. But I don't know the first thing. . . . I'm as good as a proletariat [*musancha*], as uneducated as though I never went to school. What'll I do? All those people—should I draw them a wistful picture? Ah! Ignorance is misery.[52]

It is in the radical journalism that directly addressed the problems of female factory workers that the distance between "labor" and "literature," the chasm between blue-collar women and their educated well-wishers, is most clearly delineated. Journalists wrote, it seems, not so much to bridge the distance between women laboring in factories and commentators working at typewriters, but to make the former into objects of pity, to rouse readers' indignation to act *on their behalf*.

The attention given to describing the minutiae of factory conditions and the sensuous detail of poverty indicate how foreign, even exotic, the lives of factory girls were to readers and subscribers.[53] Reporters who wrote about factory conditions, such as the "lady journalist" whose description opens this chapter, did much to form the image of the factory girl as helpless and exploited, the lonely prey of colonial capitalism and subject to all of its abuses. Graphic depictions of working conditions dotted newspapers throughout the colony.[54] Such reports tarnished the reputation of factories in the colony. An article in *Chosŏn Chungang Ilbo* (Chosŏn Central Daily), reporting in the mid-1930s, stated that "it has became commonplace for families to presume the worst of factories."[55]

Socially engaged literature and working-class radicalism were two features of the 1920s and early to mid-1930s, both a product of "colonial modernity" in Korea and a critique of it. When working-class women opposed low wages, harassment, and unsafe working conditions in their strikes and appeals to a reading public, they were claiming the instruments of capitalist modernity to condemn the industrial culture it had created. Their language was a mixture of the old and the new. They harked back to their idyllic, feudal ties at the same time that they claimed a new public role for themselves. The sparse and fragmentary nature of factory girls' writings to newspapers and magazines was an outcome of the historical suppression of the writing of lower-class females in Korea.

The suppression of these women's writing was, of course, related to the economy of literacy in colonial Korea—many laboring women were illiterate. According to a 1922 government survey, 59 percent of waged workers had reportedly never attended school.[56] The figure for women alone was much higher. According to the Korean Government General Census Report of 1930, over 92 percent of Korean females were completely illiterate.[57] And an inspection of the magazines published in colonial Korea also indicates that older, nineteenth-century literary structures were still in place, so that *kisaeng,* female entertainers or courtesans who were part of the Chosŏn dynasty caste system, were able to bring out their own magazine, but "modern" factory girls never did.

GENDER PANIC

The aforementioned Kyŏngsŏng Rubber Factory strike marked the beginning of a long string of factory girl strikes that pitted these workers against their employers and the police, and brought them publicity and a new distinction in Korea's burgeoning capitalist society. After the 1923 strike, rubber factories became the site for some of the fiercest industrial campaigns of the colonial era: in August 1933, in P'yŏngyang, almost a thousand rubber workers from different factories went on strike over reduced wages.[58] The action of the women at the Kyŏngsŏng Rubber Factory was significant, not only for the support and publicity it gained, but also for the grievances that it aired. The employees had stopped work to protest discrimination in wages between male, female, and child workers and to expose sexual harassment. These same remonstrations emerge whenever we glimpse, in newspaper reports or literature, the lives of factory women from this era. The question then becomes, What were the broader changes occurring in gender relations at this time that made the factory and other worksites the locus for gender panic and with it sexual violence?

The feminist labor historian Anna Clark has described plebeian culture in the early years of Britain's industrial revolution as being afflicted by a "sexual crisis"—as changes in the sexual division of labor upset older sexual and gender ideologies. She describes this sexual crisis as an "enormous, disturbing upheaval in gender relations [that] accompanied the transformation of the masses of working people—the artisans, small shopkeepers, laborers, laundresses, needlewomen, servants, and sailors of the eighteenth century—into a working class of wage earners."[59] Clark uses the term "sexual crisis" to describe both a shift in the gender division of labor and accompanying social and moral changes. She observes that "many women became enmeshed in the shifting moralities of early nineteenth-century working-class culture as middle-class notions of respectability overtook an older sexual freedom."[60]

In Korea, too, industrialization forever changed the old sexual divisions of labor. The expansion of capitalist industry that took off under Japanese colonialism eventually supplanted the feudal economy of Chosŏn society where the father governed the labor activity of his wife and children. And like the industrial revolution in England, Korea's capitalist industry brought greatest change to those who threw their lot in with the factory economy or the mines, the class that had to survive by working in the new industries. It was working-class women who became the focus of the conflicting gender and class ideologies that reverberated through the transition from land-based, village-centered feudal society to an industrial one. As men seemed to lose the economic control of their family, female family members took on new waged roles that brought them into the public sphere of incipient capitalism—factories, shops, brothels, offices, schools, and so on. The prospect of female economic independence, an independence that anticipated a new social autonomy, shook Korean society.

But the sexual crisis was not a simple struggle between "feudal" and "modern" gender ideologies. Rather, colonial modernity and capitalist industrialization were adding new layers of complexity to the lives of men and women. While it opened up new ways of exploiting young women, capitalist modernity also gave them a new public role. Yet these developments were fraught with contradictions. In colonial Korea, working women found that they embodied all the things that, traditionally, a woman should not be—laboring in factories far from home, losing their beauty in the pursuit of money, exposed to all the menace of public employment, as well as being poor and without protection. Their only virtue was that they were often working on behalf of others, their family. Factory girls thus trod a precarious path through the industrial revolution in Korea and personified its contradictions—whether as the "victims" of a factory system that required all their strength and perseverance or as "corrupted" by the modern capitalist market that exposed them to its worst abuses.

The sexual harassment of working-class women, and other women who went out to work in this period, had two clear functions: it kept them in sex-segregated jobs, and it reinforced their subordination to men in the workplace.[61] This double subjugation has led historian So Hyŏng-sil to suggest that sexual harassment was also a deliberate management strategy. Writing about the systemic sexual control of women workers in the rubber industry and silk-weaving factories, So argues that foremen's and supervisors' strategy of entering into sexual relations with factory girls ended up providing them with the means to divide and control the workforce. She points out that in Yu Jin-o's novella *Yojikkong* (Factory Girl), discussed more fully in the next chapter, we observe how the supervisor "selected specific workers to conciliate, bribe, and get into sexual relations with in order to be able to survey the movements of the other workers."[62] Factory girls who copulated with their foreman were often also suspected of being informers and, due to this or to disapproval or jealousy of the patronage they received, were frequently shunned by their fellow workers. Thus sexual harassment could be an industrial issue also, a strategy to co-opt some factory girls and intimidate others, to create disunity among workers. At heart, it expressed the degree of control that male foremen, and male coworkers, believed they could exercise over females and shows how few rights were accorded to women when they entered the new social space of the factory.[63]

Yet despite the obstacles they faced, working-class women found ways to resist the treatment they received in the colonial economy. That they turned to industrial action and to the modern media to address their society has much to do with their historical moment, in which writers began paying a new critical attention to their own society and exponents of the new "proletarian literature" movement began to depict workers as the central revolutionary subjects of art and politics. The language that women workers themselves used to describe sexual harassment was often vague, and yet violence against factory girls was cited again and again in the list of grievances that accompanied industrial action in factories where women labored.[64] Vaguer still was the response by sympathetic observers who tried to understand these new associates in capitalist society. The tentativeness of the language, alongside the brutal scenes they depict, raises an important question about the limits of representation of the first factory girls. As though frightened of invoking the very violence alluded to, caught within the mystification through which sexual harassment functions, representation of such violence often failed to do justice to lived experience.

From the writings of Yi Sŏng-ryong and the *Kaebyŏk* poem author, it appears that the utopian society that some factory girls longed for was located in the feudal past, not the capitalist or socialist future. Both writers hark back to a patriarchal society that, though oppressive, expounded the protection of women, and they use that model to criticize the industrial culture that exploited them ruthlessly. This ambivalent voice captures the industrial moment of the late 1920s and early 1930s and is related to women workers' ambivalent social position. In a society that viewed female labor outside the home as a sign of subjugation, young women laboring in factories never quite achieved social sanction.[65] The repressive dormitory system, which has been compared to minimum-security prisons, was the factory's response to workers' need for "protection," promising to guard young women and their sexual reputations from the outside world.[66]

The 1920s and 1930s are often represented as a historical moment when Korean women first began to encounter modernity through access to modern employment society. Indeed women's emancipation by modernity was seen to hinge upon their entrée into modern forms of labor: in schools, hospitals, factories, public transport, art galleries, telephone exchanges, and so on. We might go a step further and suggest that labor itself—the idea of doing work that equally absorbs and rewards—emerged as a newly valorized category for men and women in this emerging capitalist society. It had a gendered aspect also: the idea that women might find liberation or "selfhood" through labor was a belief that prominent so-called New Women, as well as socialist women, endeavored to substantiate through their own life choices. But how was one to find selfhood in the fortress factories? The *Tonga Ilbo*'s 1929 column for and about working women is instructive here. Under the contingent heading "Becoming Career Women," all the trials of working widows, *kisaeng*, and orphaned factory girls were detailed and exposed. In this publishing act that invoked the new subjects of colonial capitalism, the very vessels of gender panic, here we encounter the profound pessimism about the female proletarian.

WOMEN IN THE LABOR MOVEMENT

If the social position of working-class women was dubious, so too was their political status. Working-class women in the 1920s and 1930s who protested their conditions were endeavoring to carve out a political space for themselves in a labor movement whose leadership was dominated by men. The feminist publication *Han'guk Yosongsa—Kundaepyon* (A History of Korean Women—The Modern Period) stresses how this was to the detriment of the labor movement: "Labor unions did not actively take up the causes of women workers and did not give attention to the efforts to organize women workers. As a result, women workers were not drawn into the rank and file of union organizations in large numbers, and the level of organization of female workers stayed very low."[67]

The feminist historian Lee Hyo-chae, in her meticulous study published in 1978, was the first to restore early twentieth-century female proletarian leaders, writers, and socialists to their rightful place in Korean history.[68] She chronicled the major industrial campaigns at female-dominated factories and workplaces, strikes that have been overshadowed by the larger and better-organized disputes like the Yŏnghŭng strike of 1928 and the 1929 Wonsan general strike, which for many historians had traditionally defined the potential of the colonial labor movement.[69]

Lee's work was taken up and expanded by the labor historian Yi Ok-ji. Yi warns how easy it is to overlook the presence of female proletarians in industrial disputes where workers are referred to by the generic *chikkong* (factory hand), a gender-neutral term that presumes the dominant gender.[70] The participation of women workers in strikes alongside male colleagues can be detected in such information as the demands issued by the factory workers, Yi tells us, such as those proclaimed by employees of Sunota Clothing Company during a 1926 strike where, among the grievances, was "the sexual assault of factory girls by the Japanese supervisor." From this information Yi suggests that women workers probably took part in the strike.[71]

The years from the 1920s to the mid-1930s were the most active period of Korea's colonial labor movement. By the late 1930s, however, once the labor movement had gone underground and many of its members were in prison, female factory workers found it very difficult to organize strikes or campaigns in the big factories that had been the centers of labor disputes. To locate the factory girls' response to their oppressive circumstances in the large factories of the colonial war economy, Yi Ok-ji urges us to look closely at fragmentary newspaper reports that describe factories in this period as edifices resembling prisons. As evidence of how women protested factory work, she points to tales of young women fleeing their factories being rounded up at local railway stations and returned to their employers.[72]

Even Chŏnpyŏng, the revolutionary labor organization that sprang into existence after the Japanese surrender in 1945 and seemed to offer the prospect of worker control of the factories, on closer examination appears to have been addressing an exclusively male working-class constituency.[73] Although women made up 25 percent of the reputed 500,000-strong membership of Chŏnpyŏng in 1946, Yi Ok-ji finds no evidence that women held a single official position in the organization, either in the national leadership or in the branches.[74] In fact, the only references to women workers that Yi is able to locate are in propaganda material publicizing "factory girl strikes" at silk-reeling and cotton-spinning factories.[75] Even in the organizations that purported to be seeking their emancipation, women appeared doomed to take on the role of victim, passive supplicants rather than people capable of representing themselves.

CONCLUSION

The question of where factory girls fit into Korea's colonial modernity has long been a compelling one for feminist historians. While some young women were encountering colonial modernity in the new public spaces of schools for girls, cafes, friendship societies, and the department stores of the metropolis, lower-class females were being modernized in the colony's modern brothels and factories. They learned not only the time discipline rule that tied them to whistles and clocks but also respect for machines that could crush them[76] and how much value could be extracted from their labor. The learned that some modern pastimes required a salary and leisure they did not possess. They learned, in short, that "the most feudal system of authority can survive at the heart of the most modern of factories" and that this was one of the many contradictions of modernity.[77]

In their published writings, the insights that factory girls brought to their society were many and varied. These women were some of the first to grasp how brutal "modernity," in the shape of modern capitalist relations, could be. They shared their dismay at the sexual coercion that appears to have been part of factory culture at many workplaces, and they communicated their belief in the possibility of change through industrial action and through public enlightenment. But how little they wrote, how fleetingly their own words appeared in the pages of newspapers and magazines—that is also part of their story. At a time when educated women of the middle classes were just beginning to claim a portion of the literary world for themselves, how distant was literature to factory girls, the daughters of peasants and slaves, their lives "silenced by ordinary labor."[78]

The knowledge that working women gained of their society spills forth from the fragments of writing quoted in this chapter, which are almost all that remain of the published lives of factory girls from the colonial period. Factory girls in this period symbolize key social anxieties of their era: the breakdown of the family economy, the modernization of women in the new social spaces of the factory and the modern brothel, and the colonial subjugation of Korean workers. In these years, factory girls were sometimes depicted by newspaper journalists and editors as representing the poverty and desperation of rural families, a sign not of economic advancement but of economic crisis.[79] To engage in outside work that separated you from the protection of village and family, that tossed you anonymous into the labor market, was to embody the desperation of the times.

Before closing this chapter, I want to turn again to the paradox of the first "modern" factory girls in Korea, who were so crucial to colonial industrialization though their voices were almost completely lost in illiteracy and through working conditions that afforded them few opportunities to express themselves in writing. Using only the traces they left behind—a poem, a letter, the strike demands of working women, and the reports of visiting journalists—I have tried to piece together some of the fragments of their experience. But factory girls themselves never entirely disappear into the structures that rely on their suppression, and here we must acknowledge the modern printing and publishing industries of early capitalism in Korea. The capitalism that produced a new social division of labor and destroyed the old ways also brought a new access to the means of literary production, what cultural historian Michael Denning has described as "the proletarianization of writing."[80] No longer solely the province of men and women of independent means, literary production, along with schools, experienced a boom in twentieth-century Korea, opening the reading and writing of literature to female proletarians.[81]

In the fragments of literature and journalism in which the factory girl appears, something of the nightmare of her labors is revealed to us. But that she appears only in fragments tells us of the distance between her and literature. It was a distance that some authors in the colonial period tried to broach, but the attempt to represent working-class women in proletarian literature was not without difficulties. The next chapter examines how factory girls were represented in four works of proletarian literature that reveal her at the very center of contradictory class and gender ideologies. I conclude this chapter by turning to another piece of literature, a poem by the Kang Kyŏng-ae that appeared in the magazine *Shin Yŏsŏng* after Kang had spent six months working in the factory districts of Inch'ŏn. The poem serves here as something of a bookend, along with *Kaebyŏk's* anonymous "Factory Girl," published five years earlier, and illustrates that in the intervening years a response to the appeals of factory girls was building in the new proletarian literature movement.

REPLY TO A BROTHER'S LETTER

Brother!
At last your letter has come
Asking "How is my darling sister?"
My brother!
After you were taken away your sister[82]
Never again wore silk ribbons in her hair,
And sobbed with her fist in her mouth when the rice jar was empty.
Brother, I am no longer that little girl.
That foolish weeping baby.
Now I make sturdy rubber shoes in the factory
Ah brother, you should see my forearms
They are larger and stronger than yours.
I am no longer the sister who yesterday sat on your knee eating sweets.
The year is nearly over
The wind whirls through the streets.
Darling brother, do you know? Have you heard yet?
Be happy for your little sister
Who no longer shrinks modestly but stands tall.
I am the guide of all the girls in the factory
And I *** with the factory owner till the blood rushes to my face.[83]

In the next chapter we meet a new decade of factory girls, equipped at last to overturn the sexual violence in the factories.

2. Tales of Seduction

You must not tell anyone back home what has become of me.
CH'AE MANSHIK, "P'alyŏkan Mom" (Sold Body)

In August 1933, the noted fiction writer Ch'ae Manshik published a short story about a young rural man who goes searching for his sweetheart in the factories of Seoul. He is unable to find her there but soon learns of her fate. She was deceived, sold to a brothel by a recruiter masquerading as a factory agent. Duped by forces beyond their ken, these young people must learn the lesson Ch'ae Manshik would have his readers study also—the brutality of modern capitalist relations. And who better to instruct in this message than the factory girl—young, unlettered, alone—these three ingredients sum up her appeal to unscrupulous recruiters and compassionate readers alike. Yet Ch'ae does not end his story there. The young man, Kyŏnu, ventures into the prostitution quarter to look for Jinyŏ, the young woman. After days of searching he finds her, displayed quite openly in one of the brothels. The brothel proprietress reads him in a glance and gives him an hour with his sweetheart—the worst whore in the house. Here in the parting scene, Ch'ae's realism displays its pathos:

> At the entrance to the brothel Jinyŏ wept, her arms around Kyŏnu's neck.
> "Please come back and find me. Come back next year."
> "Of course I will."
> Big tears spilled down Kyŏnu's face.
> "And you must not tell anyone back home what has become of me."
> "Of course I won't. . . . Next year I'll come back with the money to buy you out of this."
> "Where will you get that kind of money? This is my life now."
> "Don't say that . . . if not next year, then the year after. And if not then, then the year after that, but I'll bring it."
> "Just promise me you will come back and see me next year."
> "Of course I will."

"Promise me."
"I promise."[1]

They part and here we have a love story that goes some way to explaining the meaning of bonded labor and conferring interiority on those indentured by it. Concerned not so much with the deception, but with how one lives with the consequences of the duplicitous female labor market that sells bodies out of their owners' possession, Ch'ae Manshik here presents a love story very much of its time. Yet rather than hover over the transaction that distinguishes the innocent factory girl from the sullied one, the great Ch'ae Manshik is concerned with how life, and love, goes on in this economy. The fidelity of the young lovers matches the steadfastness of the debt that can never be repaid, not even with a lifetime of labor. In the lovers' pledge to love each other still, to cherish each other in the most brutal of circumstances, Ch'ae presents his readers with a compelling dilemma: is there a meeting point at which some people will refuse to be sold and others will refuse to buy them? This was the question at the heart of capital-labor relations in the colony.

This chapter examines how it came about that Korean writers in the early 1930s chose factory girls as the mark of their critique of capitalist market relations. They depicted factory girls as the sexual victims of capitalism, and in so doing they confronted the violence at the heart of Japan's colonial industrializing project. But how was interiority—agency, desire, conflict—to be conveyed in a literary form that threatened to overwhelm the individual proletarian? How was proletarian literature related to what was really happening in the factories? And how was female subjectivity to emerge from a literary movement concerned with the factory girl at her most abject and degraded?

PROLETARIAN LITERATURE

"That these writers could even contemplate the poor as central, individualised characters establishes the period," writes Sheila Smith, speaking of the transformation of the novel in England into "serious literature" in the 1840s and 1850s as writers began to imaginatively depict the inhabitants of the slums and factory towns, thereby paying a new critical attention to their own society.[2] The same radical impulse might be said to have motivated Korea's literary modernism from the mid-1920s, committed to realizing in literature the terrible costs of colonial industrialization. Like many other trends taken up in the colonial period, proletarian literature was a foreign import, adapted to suit the distinctive terrain of colonial Korean society. It was Korean students in Japan, reading Japanese proletarian literature journals such as *Tane Maku Hito* (Sower) and *Bungei Sensen* (Literary Frontier), who would start up a proletarian literary movement when they returned to Korea.[3]

In July 1925 the Proletariat Literature and Art Movement Association was formed with an initial emphasis on literature, but it soon widened to encompass plays, film, music, and the fine arts. Proletarian fiction first appeared in Korea in the mid-1920s in the magazines *Kaebyŏk* (Creation),[4] *Chosŏn Chikwang* (Light of Korea),[5] *Sin'gyedan* (New World),[6] and *Chosŏn Mundan* (Korean Literary World)[7] or as novels and novellas serialized in the *Chosŏn Ilbo* and the *Tonga Ilbo*.[8] While letters and journalists' views appeared in the newspapers, poetry and short fiction was being sent to magazines from people who would be hard-pressed to describe themselves with the neat tag of "writer" or "worker." The writer Kang Kyŏng-ae, for example, was a school teacher in her native town of Changyŏn, but she worked in the factories in Inch'ŏn before migrating to Kando with Chang Ha-il in 1931. The conventions of proletarian literature inform the world of her 1934 novel, *In'gan Munje* (The Human Predicament), where rural proletarians become urban workers, servants become factory girls, and factory girls become militants.

The Korean proletarian literature movement, which converged around the Korean Proletarian Artists Federation (KAPF), in existence from 1925 to 1935, drew artists, poets, intellectuals, and novelists together as part of a movement for the cultural enfranchisement of the working class, fashioning workers as the central subject of art and politics. In much of proletarian literature, working-class heroines are sexually harassed or raped by men from the managerial or capitalist class, images that reinforced the mandate of the male-dominated labor movement to exert themselves on factory girls' behalf.[9] But even as writers of proletarian literature addressed sexual violence in the factories, and showed how factory managers attempted complete control over the lives of female employees, authors often displayed women as the erotic object of this violence. I argue that this representation of factory girls as the sexual victims of industrialization did not furnish them with a political voice; instead it cleared the way for them to occupy the "feminized position of victim."[10]

In an examination of four factory girl stories I show how the authors of proletarian literature ran into difficulty when they attempted to represent working-class women. The stories by KAPF authors and sympathizers of the harassment and rape of factory girls, and their rescue by radical males, tells us a great deal about the construction of a socialist masculinity in colonial Korea. Employing a trope of seduction, they suggest that working women were passive in the face of sexual attack, and their only hope for deliverance or retaliation lay in a masculinized socialist politics. I suggest that while these tales offer important clues to the *prevalence* of sexual violence in the factories in the years they were written—violence that was often hard to substantiate through conventional sources—they offer little, if any, possibility of freedom from violence for working-class women. Yet other authors, too, not directly associated with the proletarian literature movement but influenced by its challenge, took to writing factory novels. Both Ch'ae Manshik and Kang Kyŏng-ae, two of the finest modern writers of colonial Korea, turned their attention to working-class characters in the new ports and factories of Seoul and Inch'ŏn. Deeply influenced by the challenge of the proletarian literature movement that in the early 1930s swept the globe,[11] Kang wrote of factory girls as the people best equipped to combat the culture of sexual violence in the factories. Her 1934 novel, *The Human Predicament,* presented greater possibilities for working-class women, both as fully individualized literary characters and as thinking and active political beings.

The question of how sexual violence became framed as a trope of seduction is a crucial one. "Seduction," because it is an ambiguous term, evokes the complexity of power and sexuality that existed in factory conditions, where the factory hierarchy often mediated the sexual lives of its employees, including mediating the nature of consent. And I use the word "trope" because in proletarian literature the seduction of factory girls referred not only to the violence in the factories but also expressed anxieties about class exploitation, about women's new working roles, and about the upheaval of industrialization carried out by a coercive, imperialist state. A number of literary and social historians have looked at seduction narratives in early capitalist societies as a political metaphor for class conflict, most notably Terry Eagleton and Anna Clark for England and Michael Denning for North America.[12] In making the connection between "lived experience" and literary trope, I am particularly indebted to the work of Anna Clark, who has argued that popular tales of the seduction of servants and factory girls by aristocrats and mill owners in late eighteenth-century England expressed both the crisis in gender roles brought on by a modern waged labor market and the trauma of industrialization for the traditional family economy.

In colonial Korea, the seduction narratives that featured factory girls had an added significance: by addressing itself to a factory system that normalized sexual harassment and rape, Korean proletarian literature paid a crucial attention to the working lives of blue-collar women. But, as we will see, using a trope of seduction was not without its problems. The ambiguity of seduction—the ways in which on the one hand it can set up a passive female subject who needs to be either rescued or "taught" what she desires, or more fruitfully can elaborate the social context for the forced acquiescence of women to their employers—makes it a valuable concept with which to read more closely the desperate politics of collaboration in colonial-era factories.[13] In Korean proletarian literature, factory girls are always in sexual danger, and the fusing of seduction, rape, and complicity alerts readers to the complexities of how power is enmeshed in employer-employee relations.

When members and supporters of KAPF began to represent working-class women in literature, they use tales of the lives of factory girls to critique their society. In Yu Jin-o's short story "Yŏjikkong" (The Textile Factory Girl), serialized in the *Chosŏn Ilbo* in January 1931, and in Yi Puk-myŏng's "Yŏgong" (Factory Girl), published in *Sin'gyedan* in March 1933, working-class women are portrayed as inherently vulnerable.[14] These two male authors of proletarian literature have their working-class heroines rescued by radical working-class males. Depicted as handmaids to revolution, the working-class women in these stories come to political awareness, not through their own experience in the factories, but via the careful instruction of male colleagues. These factory girls, portrayed as incapable of effecting their own emancipation, alert us to how limited and marginal were the roles available to working-class women in the political and literary movements that aspired to liberate them.

Like the radical journalists and activists in the male-dominated labor movement who took up the "cause" of working women, Yi Puk-myŏng and Yu Jin-o concentrated on the apparent helplessness of factory women and failed to present working-class women as capable of defending themselves or achieving their own emancipation. In the works of these two authors, the sexual crisis in the factories faced by working women was rendered as the trope of seduction. The gender roles made slightly fluid by political-economic changes could not be dealt with except in terms of a new aesthetic of victimization, which reestablished the passivity of females.

TALES OF SEDUCTION

In Yi Puk-myŏng's "Yŏgong" (Factory Girl), a short story published in 1933, the factory girl heroine Chŏng-hŭi is the romantic partner of the radical male worker Ch'ang-su, his sexual reward for going to prison or otherwise taking on the burden of the class struggle. When Ch'ang-su teaches the unsophisticated Chŏng-hŭi the alphabet in English, he also instructs her on the proper duties of women and gives her lessons on "comradely love."[15] After Ch'ang-su's imprisonment, Chŏng-hŭi is pursued by the factory overseer, who presses money upon her and tries to kiss her. When she resists, the Japanese overseer blames the influence of Ch'ang-su, turning the harassment of Chŏng-hŭi into a battle between two men. Although Chŏng-hŭi is the main character in "Factory Girl," she is not the principal agent. That responsibility is given to Ch'ang-su, whose influence and leadership, even though he is absent for most of the story, directs the workers/ protagonists who blossom into class consciousness when they enact his teachings and begin a strike.

Yu Jin-o's "The Textile Factory Girl" is a longer and more complex work that found a wider audience when it was serialized in one of the colony's leading daily newspapers in 1931. The main protagonist is Oksun, a young unmarried woman who works in a silk-reeling factory.[16] Oksun's father is disabled from a fall on a construction site and Oksun and her mother must work to support their family. Oksun is not interested in politics but she admires her old friend Kŭnju, a factory girl not much older than Oksun but married and with children, who speaks out openly against the managers' plan to cut wages. One day Oksun is called into the foreman's office and there she meets the Japanese-speaking director of the factory, Chŏnjung.[17] Chŏnjung compliments Oksun and gives her a present of money, saying that he keeps an eye out for the poor girls. Unable to refuse without annoying this powerful man, Oksun also agrees to insinuate herself into Kŭnju's confidence, and the next evening she visits Kŭnju's home, a sweltering room located under the railway line.

There, Oksun—and the reader—discover a socialist cell led by another radical male, Kŭnju's husband Kang-hun. Oksun is deeply impressed by the young workers she meets that night, and she is surprised to see some of her acquaintances from the factory, women whom she did not realize were politically active. But Oksun is uncomfortable and nervous during the meeting, and the next day, when she is called into the foreman's office again, she discloses to Director Chŏnjung the identities of those in the cell. Chŏnjung is delighted with her information, but Oksun's fear is mounting. Chŏnjung asks the Korean foreman to leave the room and then he rapes Oksun. Yet what starts off as a description of a violent rape ends as a narration of a sex scene. In this important passage, the author steps out of his identification with Oksun in order to describe, as though from an objective distance, the Japanese director having sex with Oksun: "Their two bodies fell with a thud onto the worn linoleum floor. When Oksun recovered her senses she was no longer a virgin. . . . She looked up and saw the sweaty face of the director as he sat smoking a cigarette. . . . "You're okay?" he asked."[18]

As rape is elided into sex, Oksun becomes an erotic object both for the factory manager and the author himself. At the end of the story, when everyone has been sacked and Oksun has turned to radical politics, the author concludes: "And so the company, after shaking off the undesirable elements in the factory (although among them the director was able to taste one unexpected honey sweet), completed its third stage of restructuring."[19] As though torn between the merits of a politics of emancipation, and an appreciation of factory girls just as they are—as exploitable "honey sweets"—Yu Jin-o offers us this disturbing story where rape becomes the trigger for a factory girl's political awakening and violence against her is eroticized.

Political authority resides in the men in both these stories, and while Kŭnju is the most knowledgeable person on the factory floor it is her husband Kang-hun who must teach her and the others about the theories that inform their actions. While the radical men in both these tales, Ch'ang-su and Kang-hun, are impervious to the temptations of betrayal, the factory girls teeter dangerously close to it and, in Oksun's case, "succumb." It appears that these women's sexual virtue is being tested as much as their politics. Yet neither Chŏng-hŭi nor Oksun are able to combat the sexual harassment they face in a way that recognizes it as a problem that factory girls themselves can confront. It is on this point, and in writing factory girls as the principal agents of their own "awakening," that *The Human Predicament* allows working-class women a greater political imagination.

THE HUMAN PREDICAMENT

On the first of August 1934, the first installment of a new novel, *In'gan Munje* (The Human Predicament), was published in the pages of the *Tonga Ilbo*. The novel's 120 chapters concluded on December 22 of the same year with the death from tuberculosis of its principal character, Sŏnbi, a factory girl. Set in the early 1930s, the novel vividly portrayed the lives of young country girls and their foray into metropolitan life, following them and their friends through Seoul's jazz cafés and secret revolutionary societies, the fortress factories of Inch'ŏn, and finally to death, in the textile factories. The novel's author, Kang Kyŏng-ae, was twenty-eight years old at the time of its publication and resided in provincial Kando (Jiando), in southeast Manchuria.[20] Kang had already made a name for herself as a poet, essayist, and writer of short fiction in the magazines *Shin Kajong* (New Family) and *Shin Donga* (New Asia). With the serialization of *The Human Predicament* in the *Tonga Ilbo,* she made her reputation.

The Human Predicament is a fictional account of the consequences of Korea's rural destitution of the 1920s and 1930s for young women and girls. Kang wrote about the painful, compromised, and incomplete journey from feudal village to modern factory. In making factory girls and concubines the center of her major work, Kang explored what the modern world had to offer impoverished female colonial subjects. Part of her originality was her very closeness to the world of the characters she portrays, and it should therefore come as no surprise to her readers that her early life was shaped by poverty.

Kang Kyŏng-ae (1906–44) was born in Hwang-hae Province in 1906.[21] Her father was a farm laborer, and after he died in 1909 her mother married a man in his sixties who had children of his own but who was able to financially support the widow and her daughter. Like the daughters of progressive, well-to-do families, Kang entered the modern school system for girls,[22] however her family was neither tolerant nor wealthy. When her mother had remarried, she had entreated her new husband's family to enroll her daughter in school, and in 1915 when Kang Kyŏng-ae was nine years old she began at Changyŏn Elementary School for Girls. But despite Kang's thirst for learning, the family appears to have been disinclined to spend money on her, or on her mother, who became a virtual servant in the house of her new husband. Throughout her school years, even at the well-known Sunghŭi Girls School in P'yŏngyang, Kang had trouble paying her school fees, and several sources record that she was punished at school for stealing money and other things from her classmates.[23] Her own writings reveal a child ostracized and humiliated at school and constantly in fights with her older stepsiblings at home.[24]

In 1923 Kang took part in a strike at Sunghŭi Girls School, a missionary school run by an American headmistress, an action that attracted considerable publicity.[25] The strike took place in October 1923 when Kang was in her third year of middle school. She and several other boarders were expelled for protesting draconian conditions in the dormitory when, at the Chusŏk holiday, a student was refused permission to visit the grave of her friend. At this time, Sunghŭi Girls High School was derisively referred to by some as P'yŏngyang's second penitentiary.[26] Kang's political instincts appear to have been honed early, a not uncommon occurrence in the middle years of the colonial period when many Korean schools were hothouses of political activity.[27]

Kang Kyŏng-ae is an author who has eluded definition. Her institutional affiliations are tenuous—although she wrote proletarian literature she never joined the Korean Artists Proletarian Federation, and her involvement with the feminist Rose of Sharon Association (Kŭnuhoe) was marginal.[28] Her output is eclectic—poetry, short stories, essays, book reviews, two novels, and a host of autobiographical writings. Her novel *The Human Predicament* barely survived the tumultuous era in which it was published. After her death in 1944, and the collapse of the Japanese empire in East Asia in 1945, Kang's husband, Chang Ha-il, went to North Korea, where he became associate editor at *Nodong Sinmun* (The Daily Worker) newspaper. Yi Sang-gyŏng, a leading authority on Kang Kyŏng-ae, surmises that Chang is the person responsible for arranging the republication of *The Human Predicament* in North Korea in 1949 by *The Daily Worker*, thus ensuring the manuscript's survival into postcolonial Korea, north and south.[29]

Kang Kyŏng-ae wrote from a close understanding of the pressures and traps created by female poverty. A number of literary historians have noted the connection between Kang's personal experience of poverty and the destitute characters she brings to life in her fiction, explaining her compulsion for vivid social realism in personal terms.[30] Kang's realism does not only apply to peasants and factory workers, however. One of her most famous short stories, "Ŏdum" (Darkness) is about a Korean nurse in a hospital in Manchuria who, over the course of a day, unravels as the time for the execution of her communist brother draws near. In *The Human Predicament,* the author's most sympathetic characters are the girls Sŏnbi and Kannani, who must battle to gain entry into the shabbiest corners of the expensive "modern life" in the colony.

One of Kang Kyŏng-ae's innovations was to suggest that working-class women were the exemplary New Women of colonial modernity, struggling to realize some of the promises of modernity—such as greater political and personal freedoms—and to overcome some of its more brutal manifestations. The heroine of *The Human Predicament* is Sŏnbi, the orphaned daughter of poor peasants, who is taken in as a servant by the village landlord Tŏkho, is raped by him, and then flees to seek work and anonymity in the factories of Seoul and Inch'ŏn. Sŏnbi seeks the friendship of Kannani, another village girl who was Tŏkho's concubine before Sŏnbi. Kannani has made a new and independent life for herself in Seoul. It is Kannani who teaches Sŏnbi how to survive in the metropolis and obtains jobs for them both in a new textile factory just opened in Inch'ŏn.

Sŏnbi's traumatic journey is narrated alongside that of the main male protagonist, Ch'ŏtchae, a village boy who lives with his mother in bitter poverty.[31] Ch'ŏtchae's mother is the sexual instrument of the village men, and Ch'ŏtchae grows into an angry youth, given to drunkenness and fights. But he possesses a fierce sense of justice, and when he and the other villagers are swindled by Tŏkho he is determined to study "the law" that has robbed them. Ch'ŏtchae makes his way to the docks at Inch'ŏn, where he meets the university student Sinch'ŏl and discovers the labor movement.

NEW WOMEN AND SUPERFLUOUS MEN

By contrasting the three main protagonists of the novel (Sŏnbi, Ch'ŏtchae, and Kannani) with their peers from a higher class (Tŏkho's daughter Okjŏm and the university student she idolizes, Sinch'ŏl), the author is able to illuminate how poverty and social class mediate the experience of capitalist modernity in colonial Korea. It is when she turns her attention to the wealthy New Woman, Tŏkho's daughter Okjŏm, that Kang Kyŏng-ae calls into question the possibility of female solidarity in colonial Korean society. The term "New Woman" was coined to describe those women who adopted modern ideas and practices as a result of acquiring a modern education, an expensive accoutrements in those years. In Okjŏm, Kang Kyŏng-ae gives us a New Woman full of spite and greed. As the social historian Kang Yi-su puts it, "As the daughter of the wealthiest landowner in the district, Okjŏm, the college-educated 'New Woman,' has no contact with the servitude and oppressiveness of the colonial condition; only the problems of free love and 'desire' concern her, and as such she exemplifies the women of the leisured class and their dilemmas."[32]

In Kang Kyŏng-ae's novel, the promising New Women are instead those being modernized in the new social space of the factory. Kannani, who survives the factory system that kills Sŏnbi, ends up being one of the characters best equipped to recognize, and perhaps one day overcome, the predicament she shares with other Korean females of her class. Kannani understands early on that the traps of female poverty—rape and concubinage in her village and sexual harassment in the factory districts—are something she shares with other lower-class women.

In her portrait of the jealous first wife of Tŏkho (Okjŏm's mother), and of Okjŏm herself, Kang Kyŏng-ae seems to imply that there is little difference between New Women of the prosperous classes and their old-fashioned mothers. They are both shown trampling on other women, their female servants, whose domestic labor frees them to pursue self-fulfillment

(Okjŏm) or to enjoy domestic tyranny (Okjŏm's mother). In writing about the collective and self-emancipation of factory girls, Kang enlightens us as to the limited possibilities of solidarity between women from different classes in colonial Korean society. In her novel, working-class women and servant girls are "victimized by all other women,"[33] and Kang implicates bourgeois women in her critique of the society that exploits village girls.

But Kang reserves her deepest critique for the intellectual hero, the "superfluous man," the solitary and idealistic university student Sinch'ŏl. The superfluous man is the creation of the Russian writer Turgenev, whose protagonists are shown caught between idealism and a repressive state, men who read Rousseau on their landed estates and dream of change.[34] When Korean left-wing intellectuals in the 1920s and 1930s compared their country to the state of Russia fifty years previously, they also borrowed the terminology of Russian writers and adopted their fictional characters, like Turgenev's superfluous man.[35] Turgenev and other Russian writers' ambivalent portrayal of intellectuals, aristocrats, and clerks, of anyone who talked of changing society but was unable to act, appealed to Korean authors in the 1920s and 1930s who found themselves similarly having a cultural freedom greater than their political liberty. Kang Kyŏng-ae's Sinch'ŏl embodies this ambivalence.

Sinch'ŏl is introduced to the reader as a university student friend of Okjŏm whom she brings home to stay with her family during her school holidays. Sinch'ŏl is the son of Okjŏm's teacher, a handsome Seoulite studying for his bar exam and prey to the great vice of young men of his generation—politics. Okjŏm falls in love with Sinch'ŏl, her family approves the match, and Sinch'ŏl's own father orders his son to marry the rich daughter of the usurious landlord. Only Sinch'ŏl cannot bring himself to trade his principles for a comfortable match. He is ambivalent about Okjŏm, and he finds that his growing feelings for Sŏnbi provide him with a further stimulus to join a political movement for the emancipation of the working classes. By the end of the novel, Sinch'ŏl comes to embody the vexed issue of male intellectual leadership of working-class movements.

Sinch'ŏl, with all his advantages, is not able to inspire Sŏnbi's love, and Sŏnbi finds herself increasingly drawn to Ch'ŏtchae. Sinch'ŏl's fascination with the servant Sŏnbi is introduced to the reader in the first third of the novel. Staying in the same house as a guest of Okjŏm and her parents, Sinch'ŏl becomes captivated with the beautiful Sŏnbi who washes his shirts. He muses with a naive perplexity on the distance that class puts between them: "If she were a classmate they could meet in cafés or other places but this was the first time he had found it so difficult to get to know a woman."[36] Sinch'ŏl considers Sŏnbi his equal and Okjŏm considers her a rival, only because of Sŏnbi's superior beauty. In Kang Kyŏng-ae's novel, female beauty is a commodity that crosses class boundaries, a trope familiar from classical Korean literature.[37] Though she is from a respectable village family, bereavement has robbed her of parental protection and her vulnerability is known throughout Yongyŏn Village. Her movements are watched proprietorially by more than one pair of male eyes—Tŏkho, Sinch'ŏl, and Ch'ŏtchae all dream of her.

THE RAPE OF SŎNBI

Orphaned at age seventeen, Sŏnbi is invited by Tŏkho to move into his household and take his daughter Okjŏm's bedroom while she is away at school, appearing to treat Sŏnbi as his own daughter. Sŏnbi complies and comes under Tŏkho's protection, where he takes the role of both employer and guardian. Things begin to go wrong when Okjŏm returns and unceremoniously turns Sŏnbi out of her room. The pert young Okjŏm knows a servant-class girl when she sees one, and sisterhood is not attempted again. Kang Kyŏng-ae understands the powerful fantasies surrounding the new possibilities for getting an education in the colony, however, and halfway through the novel Sŏnbi indulges in her favorite dream—going to school.

When Tŏkho makes his first sexual advances to Sŏnbi he uses the lure of education to attract her, promising that he will send her to school like his own daughter, Okjŏm. It was bait used to great affect at that time by factory owners to recruit girls who could not otherwise afford to go to school.[38] The cruelty of this deception is underlined when, following her reverie about education, Sŏnbi is raped by Tŏkho.

After the rape, Tŏkho comes regularly to Sŏnbi's room to copulate with her, and Sŏnbi is unable to do anything but comply. In Tŏkho's household the act of rape is, to use American literature scholar Hazel Carby's terms, "conceived in patriarchal terms to be sexual compliance," as Sŏnbi's role shifts from servant to concubine.[39] Kang Kyŏng-ae is able to show how compromised Sŏnbi is in this relationship, and once the old servant woman is dismissed by Tŏkho, Sŏnbi is utterly alone in the house and her thoughts twitch between ruling the household and escaping it.[40] The stakes are high: If Sŏnbi falls pregnant and gives birth to a male heir, she knows that she will become the most powerful woman in Tŏkho's establishment. However, if like Kannani and Tŏkho's earlier concubines she does not conceive, her downfall in the household is inevitable. Events overtake Sŏnbi when Okjŏm jealously accuses her of having an illicit affair with Sinch'ŏl, and a furious Tŏkho turns Sŏnbi out of his house. Sŏnbi embarks on a journey to Seoul that eventually transforms her into a factory girl.

In this portrait, Kang Kyŏng-ae gives us distinctly lower-class colonial heroines, without the adornment of sexual purity that was simply unavailable to most poverty-stricken girls on the move. Furthermore, the author does not forget the cruelty of the world that industrialization and modernization are leaving behind. In the closed world of the village where Sŏnbi becomes an orphan, she is exposed to the terrible fate of an unprotected, "unowned" female. But rather than suggest that there is any key difference in relations between the sexes under Confucian feudalism versus incipient capitalism, Kang shows Sŏnbi's continued victimization under both systems. Sŏnbi's pretty face and guileless ways are as alluring to the next men who enter her life forcefully—the factory overseers—as they were to the feudal landlord.

THE ECONOMY OF COMPLICITY

All three main characters in *The Human Predicament* (Sŏnbi, Kannani, and Ch'ŏtchae) are expelled from village life and must go to the city and become proletarians in order to be fully awakened to their shared human predicament. Sŏnbi moves in with the more worldly Kannani, who will be her guide in metropolitan Seoul. But Kannani has a secret: "Sŏnbi viewed with suspicion Kannani's habit of returning home late at night, her mind obviously on other things. And when Sŏnbi thought back to her own life in the village and how every night she was forced to submit to Tŏkho, she shuddered involuntarily. Sŏnbi began to wonder despondently if Kannani too might be committing that kind of vice."[41]

As the reader soon discovers, Kannani's absences have nothing to do with either male sexual predators or illicit romance—she is involved in a workers' movement. And like the socialists who were entering industries all over Korea at this time, the early 1930s, to create labor unions and improve working conditions, Kannani signs up for a textile factory in Inch'ŏn to earn a living while she secretly distributes political pamphlets to her coworkers. Kannani's political contact in Inch'ŏn turns out to be none other than the student Sinch'ŏl. Disowned by his father for not marrying Okjŏm and giving up politics, Sinch'ŏl throws his lot in with the labor movement. His first mission is in the industrial sector in Inch'ŏn, where he becomes friends with Ch'ŏtchae, while his second assignment is in the very factory where Kannani and Sŏnbi are engaged as factory hands.

On her first night in the factory, Kannani gazes out the window, intimidated by her isolation and the political aim she will pursue there. She looks apprehensively at the wall guarding the factory grounds, which seems an impenetrable fortress, and wonders if she will be able to find a hole through which she can crawl to meet her contact, Sinch'ŏl, on the outside, and obtain assistance with her mission.

Kannani is exploring the dark corridors, peering outside at the gates when a noise in the hall distracts her:

> In breathless apprehension, Kannani watched as a door creaked open. A factory girl was creeping noiselessly down the corridor and seemed to be making for the supervisor's night-duty office. Drawn by curiosity, Kannani followed. Outside the night office the girl stopped, hesitating before opening the door and going inside. "Who could she be?" wondered Kannani. . . . And then the memory of how she had tamely submitted [night after night] to Tŏkho's assaults flashed through Kannani's mind with such force that her back prickled with perspiration. She stood for a long moment caught in memories, then crept up to the night office and bent her ear [to the door]. Kannani could hear nothing. If there had been no important mission to carry out, how she would have pummeled the door and raised hell until the entire factory was turned upside down, in order to expose the truth to all the other factory girls.[42]

But Kannani does not pummel on the door and the reader is left dangling. What is this "truth" that Kannani hopes to expose? That factory girls have sex with supervisors? And how does one expose the "truth" of the power relations in this coupling? As Kannani herself knows, it may be one thing when one is immersed in the "night after night" of the relationship and quite another when one has found politics and friendship and a regular wage.

It is important to note that, unlike in Pak Hwa-sŏng's "The Night before Chusok," "The Textile Factory Girl," or "Factory Girl," Kang Kyŏng-ae's scene of an overseer and a factory girl does not depict rape or eroticized assault but rather furtive sex between the operative, who must bear all the danger of exposure, and her supervisor. Kang, with her experience in the factories at Inch'ŏn and her commitment to realism, portrays complicit sex rather than rape because this was in fact the most common form that extreme sexual harassment took in the factories. What other authors of proletarian literature preferred to gloss as rape, Kang understood was far more compromised and desperate. We might say that gender relations in the factories ran on an economy of complicity rather than a politics of rape.

Cases of rape were rarely prosecuted and were far less common in strike notices and newspapers than protests over "treatment" *(daewoo)*, "abuse" *(yok mŏkta)*, and "body searches" *(sasaek)*. Sometimes the complainants borrowed the language of their abusers, such as at the Sanship Silk Reeling Company where six hundred workers went on strike to demand a ten-hour day, edible food, and that male coworkers cease their "practical jokes." We might conclude that the language was vague—"practical jokes" meaning harassment—because the politics of male-female relationships in the factories was oftentimes messy. But a deeper mystification was also occurring here. A great deal of work went into keeping these associations vague, unverifiable, complicit, and constantly available. As we saw in chapter 1, the wage and bonus system, recruitment practices, the absence of any legal protection, and a prevailing discourse on the corrupting influence of factories all worked to mystify the politics of forced acquiescence. And it was the vagueness that killed forced sexual complicity as a political issue in both the labor movement and in proletarian literature, instead turning it back to be worked out within the four walls of the dormitories. The vagueness and the messiness allowed forced sexual complicity to thrive.

On her return to bed, Kannani warns the sleepy and bewildered Sŏnbi about the factory overseers, saying, "These overseers who manage us and all those behind them are one thousand times, ten thousand times more frightening men than Tŏkho."[43] Kannani does not tell Sŏnbi what she saw, but several nights later she disappears through a drainage vent in the factory compound for a few hours and the next morning flyers appear in odd places all over the factory—under bedding and stuck on walls, telling of all the events in the factory, known and secret. This is repeated every morning for days, and all kinds of topics are raised in the notices; one morning the factory girls are surprised to read a critique of the overseer's speech to them from the night before. The flyers bring the employees closer together and they begin to discuss with each other their lives inside the factory walls. A group of girls one night start to wonder about the bonus system:

> "Well girls, I don't know who brings us this paper, but what they write is true. The overseer did promise that every day somebody in the factory would win a bonus of 20 chon for their work, but who's seen this bonus? It's just an empty word."
>
> "What about Haeyŏng who always works so hard? She hasn't got any bonuses. . . . It must be a lie."
>
> "But that new girl, the pretty one in room seven, she's received a bonus."
>
> "She has? Who?"
>
> "Be careful what you say, someone might hear."

"Who is going to hear and who's listening anyway at this time of night?"

"Don't you know anything? Our overseer does the rounds every night."

"No, not her. Haeyŏng said it's the new girl who stands in front of her. The one our overseer always trails after smiling from ear to ear. What a disgrace. I don't even like looking at them. He did the same thing to Yong-nyŏ."

"But she's better looking than Yong-nyŏ. That's what happens with pretty girls. If I was a man I'd like her too. Her eyes and nose are so pretty."

At this a factory girl who was partly deaf piped up, "Pretty, what's pretty? There's something wrong with her hands. I'm frightened whenever I look at her hands." [44]

"Oh this stone deaf silly. What did she hear? Hee, hee, ha, ha." Everyone burst out laughing. . . .

The deaf girl suddenly threw off her quilt and, taking a deep breath, began to speak with the urgency of someone unburdening themselves:

"Listen to me. I used to work in a milkhouse in Seoul. The overseer there was a disgusting man. He didn't pay us our proper wages so why shouldn't we go out on strike? So that's what we did, only a couple of the girls snitched and told the overseer everything and everyone was fired. I was lucky not to be dismissed, but the overseer was so bad that eventually I quit anyway. That's how it ended. . . . "

"Those snitches all deserve to die! And I bet they were having an affair with him. . . . "

"Look at us. We work until we collapse and they don't even pay us our wages, what sort of a state is this? We are our parents' precious daughters and here they treat us like dirt. Today I nearly caught my hand in the machine, I could have lost it. Before we came here who among us ever thought that the factory would be like this?" As she spoke the girl put her hand to her cheek and shuddered. She seemed to hear the turning of the spinning reel.

"I'd like to meet the person who wrote this paper. Should we try to find them?"[45]

Here we can see factory girls participating in their own "awakening." The sexual economy of the factory is something they are all familiar with, and they reproduce its cruelty in their own analysis of the "seduction" of factory girls by overseers, which divides working-class women into "snitches" and hard-working factory girls. But in this scene we also perceive Kang Kyŏng-ae's stake in presenting writing as a form of political illumination. The provocative tone of the flyers (we never actually read one; we are only told about them), and their instant effect on the factory girls, intimate the inseparable links between literature and political awareness. The rhetorical message of the flyers broadens the awareness of the women to an appreciation of their shared predicament, and in this sense the flyers might be said to share, in part, one of the overall projects of *The Human Predicament*—to publicize the lives of factory girls to a society that had for so long tolerated, or not even noticed, the suppression of lower-class women.

That it should be factory girls who enact their own awakening is a distinguishing feature of Kang Kyŏng-ae's novel. As we have seen, a number of male authors of proletarian literature who wrote about factory women preferred that any political awakening be carried out by radical working-class males. The fate of the radical male in *The Human Predicament,* the student-worker Sinch'ŏl, is in sharp contrast to the exemplary role of his peers in the earlier works of proletarian literature, published before the mass arrests of 1934.

THE END OF KAPF

When in *The Human Predicament* Sinch'ŏl is arrested for his political activity in Inch'ŏn, the police offer him a choice: fit in with society and give up criticizing or trying to change it, or face prosecution as a political criminal. Deliberating in his prison cell, Sinch'ŏl is haunted by the memory of his friend Ch'ŏtchae, and we begin to realize that for Sinch'ŏl radical politics always meant fighting on behalf of others. By presenting himself as a leader—playing the part of an experienced cadre to Kannani even though he has never seen the inside of a factory—and taking the gullible Ch'ŏtchae under his wing, Sinch'ŏl has taken on a role that makes him a target for the colonial police. When Sinch'ŏl turns and gives up his political allegiances and friendships, he makes a choice that acknowledges he is not actually central to the labor movement he is involved with. For all its pathos and bitterness, Sinch'ŏl's betrayal becomes a kind of resolution in the book, for it allows Ch'ŏtchae and Kannani to come into their own as revolutionaries. Whereas in the stories of Yi Puk-myŏng and Yu Jin-o, the radical male who straddles the world of ideas and the world of the day laborer is the lynchpin of all political activity in the female-dominated textile factories, in *The Human Predicament* he is revealed as peripheral and fleeting.

Kang Kyŏng-ae's female protagonists suggest that the labor movement was not the sole province of men, and the ideas shared by Kannani and Sŏnbi underline the attraction of socialism for female proletarians. These two points are worth reiterating because, in the peculiar circumstances of the history of socialism in Korea, they have often been forgotten. The difficulties postcolonial intellectuals have had in claiming a history of socialist anticolonial resistance are most succinctly put by the sociologist Kim Dong-ch'un:

> Chosen's[46] socialism experienced an extremely unfortunate birth because it was imported from the colonizing country, Japan. Therefore we can say that there was the possibility from the beginning that Chosen's socialists, tacitly disregarding their own cultural, historical resources, could lean towards a hasty "universalism." The overwhelming majority of early socialists were those who could receive a new-style education because they were from the landlord or moneyed class. . . . For this reason, as in the case of liberalism, we can view socialism as, more than anything else, part of a new way of thinking for intellectuals.[47]

This perception of the shallowness of socialist ideas in the colony is reinforced by the fact that large numbers of professed socialists "turned" from the mid-1930s onward. Prominent among those who shifted their political allegiances were authors of proletarian literature and members of the Korea Proletarian Artists Federation.[48]

KAPF had been targeted by the Japanese police since 1931, when some of its leading members were arrested. When the poet, novelist, and critic Pak Yŏng-hŭi, a leading member of KAPF since its earliest days, left the organization in 1933, he famously said, "What was obtained was ideology, what was lost, art."[49] The move from defending and improving the status of working-class people to defending and elevating art was, in the circumstances, a predictable resolution. Because art and literature are seen by so many to transcend society and politics, loyalty to their civilizing air becomes implicitly a commitment *not* to intervene in the politics of the day, especially at a time when political intervention was becoming dangerous. As a strategic conversion for those wanting to distance themselves from any political entanglements, it makes perfect sense. But it is no less an ideological position for that.

It is perhaps paradoxical that these proletarian writers would decamp to the imaginative realm of art just when they had shown themselves incapable of the imaginative leap that would enable them to depict factory girls as fully sentient human beings. With their departure and the collapse of the proletarian literature movement, factory girls lost a powerful, if flawed, ally. Working-class women would not again gain cultural prominence in literature or the wider public culture in South Korea until the 1980s. Yet the disappointment with the perceived shallowness of the socialist tradition in Korea, such as that put forward by Kim Dong-ch'un, is usually expressed only in terms of an intelligentsia. That many working-class women also had a stake in these movements is not often acknowledged.

Kang Ju-ryŏng was perhaps the most famous female proletarian at this time to lead a labor dispute openly as a socialist. She came to prominence when she led the workers at the P'yŏngwon Rubber Factory in a hunger strike to protest low wages before going underground to assist in organizing women at other rubber factories and to provide leadership in industrial disputes. She joined the P'yŏngyang Red Labor Union and was arrested in an incident involving that union late in 1931.[50] She was languishing in prison the year *The Human Predicament* was published, in 1934. If Kang Kyŏng-ae was looking for inspiration for the character of Kannani, she could not have found a better person than the young widow who "turned upside down" the rubber factory where she was employed.[51]

THE TRAGEDY OF THE FACTORY GIRL

Kang Kyŏng-ae's ambiguous place within the male world of the proletarian literature movement—where her writing was admired for its "masculinity"[52]—her distance from metropolitan literary circles, and her conviction that money, more than education, was essential to female liberation placed her on the periphery of the organizations that purported to be seeking the emancipation of working-class women.[53] Her ambiguous position might be compared to factory girls themselves, who were at once central to the colonial economy and peripheral to the modernizing ideologies of empire and industrialization.

Kang Kyŏng-ae's achievement with *The Human Predicament* is remarkable. While Yi Puk-myŏng and Yu Jin-o were unable to fully grasp the class and gender ideologies at stake in the suppression of working-class women, and instead simplified the complex experience of factory girls into tales of seduction and rescue, Kang shows how one might appropriate the resources of the socialist imaginary for women. In her novel, Kang employs a trope of seduction, *not* as a neat political metaphor for class conflict, or to sexualize the limited agency of working-class women, but as a means to explore the politics of complicity and resistance in colonial-era factories. Her novel offers us an entry to examine the sexual violence in colonial-era factories, which journalism and strike notices often only hint at. More than that, Kang understands the ways in which sexual violence and harassment were ambiguous in this world. She shows that to focus on the violent act of rape as the sole moment of clarity mistakes the nature of sexual violence. Her characters experience the pain of complicity. Like Jinyŏ in Ch'ae Manshik's story "P'alyŏkan Mom" (Sold Body), they grapple with a new economic reality that compels new stories and new kinds of subjectivities.

The flowering of a factory girl literature was a historical and economic moment as well as a literary one. Written at a time when social realism was as much a political intervention as an aesthetic choice, this literature enables us to read gender and class minutely into Korea's industrial revolution and to trace the relations between working people and the writers and intellectuals who, in Jacques Rancière's famous phrase, had "gone out to meet them and perhaps wish[ed] to expropriate their role."[54] In a New Year's message published in the magazine *Shin Kajong* in December 1933, Kang Kyŏng-ae wrote, "If we do not see in society complete economic reform then we shall never see complete female liberation. . . . The more women are commodified, the more our standing as human beings is blighted."[55]

In fact, the capitalism that produced a new sexual division of labor also brought a new access to the means of literary production. Literary historian John Guillory has said that "the same system which 'commodifies' women in new ways also permits them to produce new commodities (such as novels), to become new kinds of cultural producers."[56] In using writing to alert her readers to the miserable living conditions of factory girls, and the need for "complete economic reform," Kang Kyŏng-ae was demonstrating that the very system of production that exploits female labor also distributes the "cultural capital" necessary to generating a critique of that system.

In writing the industrial revolution as the tragedy of the factory girl, Kang Kyŏng-ae fixed the experience of working-class women at the heart of the changes taking place in Korean society and culture in the 1930s. Part of the "productive mood" of the 1920s and early 1930s, *The Human Predicament* might easily have disappeared completely; for all its insights it, lacked one crucial element. Though it included factory girls, *The Human Predicament* did not speak to them. The high rate of illiteracy among lower-class women, and the grueling nature of their work in the factories, made reading itself a pastime beyond their compass. It would take another two generations, and a revolution in literacy, for South Korean factory girls themselves to address their society, and attempt to transform it, through literature.

3. The Road to Seoul

> I was fifteen. Palms flat against the train window I looked out at the station platform. Good-bye old home. I'm off to make my fortune.
>
> Shin Kyŏng-suk, *Oettan Bang* (The Solitary Room)

The retreat of the factory girl from literature after the 1930s until well into the 1970s was part of her departure from political life as well. When she reemerged in literature in the late 1970s and early 1980s, it was as a much more socially engaged figure, no longer cut off from society and the solidarity of other men and women. The world she inhabited in the 1920s and 1930s was the world of the factory, defined by sexual harassment and intimidation. By the 1970s, working-class women in their writings occupy a social sphere larger than their workplace. Based in tiny slum lodgings in Mansŏk-dong or Kuro-dong, they venture into metropolitan life and discover true friendship and solidarity, education and romance, as they unravel the prevailing ideologies that mediate their lives.

So too the labor literature that emerged in the 1970s and 1980s differed in a number of important respects from the proletarian literature movement of the colonial period. The early proletarian literature avant-garde was largely a literary movement, whereas the labor literature of the 1970s and 1980s was embedded in an assertive and culturally vibrant labor movement. This is borne out by the class background of the authors, where we find that the writers of proletarian literature from the colonial period came from all sections of society—a few were professional writers, some subsisted on the border between classes and wrote about the factories and mines they had labored in, and still others were "tourists" to the working class whose political inspiration was more abstract. By contrast, part of the novelty of labor literature from the 1970s and 1980s was that its creators were almost all "authentic" proletarians, many of whom wrote witness accounts of industrial disputes already famous in dissident circles.

Yet there are also striking similarities in the political conditions in which these literatures were produced. Both colonial Korea and South Korea in the 1970s and 1980s were military states, ruled by former high-ranking military commanders.[1] In both periods the state oversaw an extremely intense period of industrialization. Both these literatures flourished during a period of censorship. Indeed, proletarian literature from the colonial period was banned in South Korea under the military presidents Park Chung Hee (1961–79) and Chun Doo Hwan (1980–88), and thus the plebeian authors of South Korea's labor literature wrote their stories without being versed in the earlier proletarian literature experiment—these two literatures cannot be said to talk to each other across history.

Outside forces also influenced South Korea's path to industrialization. Like Japan in the Meiji period, Korea forced itself to undergo a version of economic development known as "rapid" or "late" industrialization. Situated far behind the developed economies of Europe, North America, and Japan, South Korea launched into a race to compete that led it to be dubbed the "catch-up economy."[2] The enormous social and personal costs of this race—the injuries, the separations, the deaths, and the degradations—are part of the history of this industrialization as much as the proliferation of jobs, the shift from kerosene to electricity, the paving of dirt roads and the building of the Seoul subway. No one appreciated these conflicting boons and costs of rapid industrialization more acutely than factory girls.

In this chapter, the factory girls themselves direct our approach to the 1970s and 1980s and make working women the foremost subject of the social world of rapid industrialization. If the representation of working-class women in literature during the colonial period revealed them as unstable figures, coded signs for what was repressed in society, it also presaged their capacity to expose and overturn those contradictions. This chapter examines the way that factory girls constructed a labor movement and a literature with themselves as subject, an act that combined literary and political representation in a radical challenge to literature and to society. Unlike earlier proletarian literature, these autobiographies are addressed to readers from both the literate laboring classes and the middle classes. Penned by autodidacts who inform their readers of their own self-directed journey into literature, these books inspired a new aesthetic form, *nodong munhak*, or "labor literature," that in the 1980s made visible the brute politics of industrial life in popular-culture form.

The very use of texts was itself the outcome of a social struggle for access to cultural literacy. When one of the authors, Chang Nam-su, describes herself in the early pages of her autobiography as a thirteen-year-old sitting high in a tree astride a branch reading *Tess of the d'Urbervilles*, her formal schooling was already over and the years of laboring by day and studying by night just beginning. When Chang spies her former school fellows winding their way home with their books and school gossip, her envy is palpable. She allows that her pleasure in reading is not unmixed, and neither is her story free from the strains of a hard-won cultural literacy.

Literacy levels among female waged workers had increased remarkably since the 1930s and 1940s. Where in 1946 nearly 40 percent of waged workers had received no schooling at all, by 1963 only 5.5 percent of workers were illiterate. By 1970, 42 percent of all workers in the manufacturing sector had acquired a secondary education, a figure that by 1980 had increased to 61.4 percent.[3] Working women were the poorest paid and usually the youngest and least educated of South Korea's modern waged workers in the 1960s and 1970s. When in the 1970s large-scale firms were required by the government to provide vocational training to their workers, textile factories instead set up on-site primary schools. This indicated the youth and neglected schooling of their largely female workforce and the irrelevance of vocational training for employees who were not in line for promotion and who lost their jobs when they married.[4]

The reputation of factory work as physically and morally harmful to women remained after colonialism ended and Japanese capitalists abandoned their interests in Korea. After then–Major General Park Chung Hee assumed power following a military coup in May 1961, he began the series of economic plans that turned South Korea into one of the world's major exporters of clothes, shoes, wigs, and—from the late 1970s—electronic goods.[5] Women, particularly young, poor, unmarried rural women were crucial to the accomplishment of Korea's economic transformation. Thus in the 1960s it was the Korean government, and not journalists or novelists, who returned to the domain of the factory, this time to extol the proper image of the working-class woman as patriotic, obedient, and frugal and to attempt to redress the disreputable image of factories and the women who worked in them.[6]

Literacy, as well as being crucial to the decision to write, also changed the relationship between white-collar and blue-collar readers. When blue-collar women began to read for themselves what was considered canonical literature in the 1970s—Leo Tolstoy, Yun Dong-ju, Hermann Hesse, Herbert Marcuse, Heinrich Heine, Thomas Hardy, Kim Sowol, Hegel, and others—what did they take from these works? Who were their literary influences, and how did this affect their decision to write? What literary historian Nancy Armstrong calls the "agency of writing" also helped confirm the role that literature played in establishing the cultural critique of Korea's capitalist society.[7] By investigating the agency of writing—literacy, literary influences, and political context—we learn much about the relationship between self, society, and culture in these years.

ARRIVAL

In the 1960s, 1970s, and 1980s, hundreds of thousands of country girls flooded into Seoul's back streets,carrying within them dreams of what they might find in the boulevards and skyscrapers of the capital. Alighting at the domed Seoul Railway Station, a shabbily grand building from Korea's colonial era, the provincials climbed the steep staircases from station platform to city center before dispersing into the shantytowns and slums of the city. Song Hyo-sun, chaperoned by her grandmother, arrived at Seoul Station one rainy night in 1973: "Watching the rain fall . . . I lost myself in the surging crowds and cars and forgot all my apprehensions."[8]

Working-class women and girls were central to the industrialization of South Korea. It is no coincidence that the factory girl reemerged as a politically charged figure in the late 1970s and early 1980s, when working-class women were at the heart of South Korea's industrialization strategy. By the mid-1980s, over one million women worked in the light-manufacturing sector, an industry with factories all over the peninsula.[9] In the areas where they predominated—the textile and garment industries, rubber footwear factories, and later in the electronics industry—the unfettered exploitation of factory girls was one of the keys to wealth creation. The economic and political marginalization of working-class women was not a tragic byproduct of the rush to modernize but was rather at the very heart of South Korea's capitalist society.

Yet, ironically, all the major players were in agreement about the value of the "sacrifice" of factory girls. In his book *Yŏgong 1970: Kŭ-nyŏdul-ŭi Pan Yŏksa* (1970 Factory Girls: A Counterhistory), Kim Won discusses that both the architects of export-oriented development and their intellectual critics shared a discourse that played on the victimization of factory girls.[10] Kim writes that, whereas for working-class men labor was an expression of economic independence and expertise, for women, particularly young unmarried women, it signified their obligation to their family. This gender division of the meaning of labor found expression within the family as well as in the labor market, so that while boys were supported as "future breadwinners" and sometimes kept away from factory work so that they might earn a college degree, their female siblings went out to work but their labor was designated only as "help for the family finances."[11] Yet buried within this rhetoric that underplayed and senimentalized the labor of a million women, a massive transformation was taking place. As Kim writes, "This discourse of the sacrifice of young women for the men in the family was used to justify the propriety of girls going off to the factories. Through the rationale of supporting the family, young women went out into the world."[12] Thus, smuggled inside the rhetoric of duty and victimization, an enormous shift in South Korea's employment society was taking place in the 1960s and 1970s.

So we must begin by fixing the marginalization of factory girls at the center of our understanding of the culture and economy of South Korea's rapid industrialization. Using their autobiographical accounts to scrutinize their social world, I argue that their foray into literature and their fight for unions to represent them were twin aspects of a struggle to emancipate themselves within a capitalist society and a wider culture that did not acknowledge them as full members. That they should turn to industrial politics and to literature has as much to do with their historical moment as their class and their own predelictions. At a time, the 1970s and 1980s, when writers and artists in South Korea were "going to the people,"[13] these young women spoke back; such was the impact of their political stand and their confident, unembroidered style that they helped enkindle the women's movement in South Korea.[14] Their influence on the labor movement and on literary and cultural forms, however, is neither so readily acknowledged nor so easy to trace.

In analyzing how autobiographical representations of working-class women expose the contradictory class and gender ideologies of industrializing South Korea, I follow the work of Patricia Johnson, who has argued that the difficulties in representing working-class women in Victorian social-problem fiction demonstrate "the masculine bias in the construction of the Victorian working class and the middle-class bias in the construction of femininity."[15] In South Korea, factory girls found themselves caught between class and gender ideologies that saw them as neither real workers nor properly feminine, and they turned to workplace politics, and to writing, to more fully represent themselves.

In South Korea the working-class autobiography came to fruition in the 1970s and early 1980s, as working-class people began to address their society, and attempt to change it, through literature. In the early 1980s, three proletarian women published autobiographies to tell their readers "a little more about the world."[16] These three women are Song Hyo-sun, whose autobiography, *Sŏulro Kanŭnkil* (The Road to Seoul), was published in 1982 and became for a time the consummate factory girl tale; Chang

Nam-su, who published *Ppaeatkin Ilt'ŏ* (The Lost Workplace) in 1984; and Sŏk Chŏng-nam, who wrote *Kongjang-ŭi Pulbitt* (Factory Lights), also published in 1984. These books and the world they illuminated found a ready readership. They appeared amid a spate of working-class autobiographies and anthologies of workers' writings, all produced through the auspices of what was becoming known as the democratic union movement *(minju nojo undong)*.[17]

What began in the early 1970s as isolated labor unrest in the garment and textile factory districts in Seoul, mushroomed into a full-fledged labor movement in the 1980s. When the regime changed following President Park Chung Hee's assassination in 1979, the new military president, Chun Doo Hwan, exerted "power without hegemony" (in Gramscian parlance) and in the midst of a repressive political climate dissident groups of students and working people grew. When the Korean Congress of Trade Unions, or Chŏnnohyŏp, was established in January 1990 as the central organization of the democratic labor movement, a left-wing cultural market was already flourishing: "In the course of the desperate struggle to build Chŏnnohyŏp, plucky workers' songs came into mass circulation, and a whole variety of working-class culture—portraits and folk music [depicting] workers' struggles and longings—developed and spread."[18]

Labor historians and sociologists have used the three autobiographies listed above as important sources on the union movement of the 1970s and 1980s. Hagen Koo quotes directly from all three books to reveal "culture and consciousness" in the union movement in the 1970s.[19] Chŏng Mi-suk uses a "life histories" approach in a case study of the textile industry to explore the world of work and the union movement in the 1970s.[20] Quoting extensively from interviews and autobiographical writings, Chŏng Mi-suk draws attention to the historical conditions that facilitated solidarity between the women. But the first person to explore the subjective experiences of class described in these factory books is historian Chŏng Hyŏn-baek. In a 1985 article that influenced all later work on the 1970s union movement, Chŏng Hyŏn-baek argues for analyzing how women workers experienced class, not only in relation to the means of production, but also in their relations with the social world.[21]

In this chapter I discuss these three factory girl autobiographies as important sources on the union movement as it developed in South Korea, both for their versions of the defining disputes of the era and for the authors' subjective accounts of what came to be defined as a mass movement—the campaign for representative unions. Yet there is deep irony in the fact that the author of *The Lost Workplace*, Chang Nam-su, helped forge the discourse of the working class and created the archetype of the female worker and the union militant, when the author herself writes so ambivalently about being a *kongsuni,* the derogatory term for "factory girl":

> When I came to Seoul and became an "industrial soldier," when carting my books off to night school, or going to prison, or passing my days as a laid-off worker, I was all the time haunted by the conviction that I was an utter nobody in this society, just a tiny cog in a massive machine. This self-derision visited me regularly. Of course, that sort of thinking disappears naturally once your eyes are opened to the truth that it is we workers who are the driving force of change in society's historical progress.[22]

This disparity between the high idealism of the union movement (dominated in these years by industrial missionaries) and the social world of its members—a gap between faith and everyday life—means that these autobiographies are a revealing source of the individual voices in a political movement celebrated for its collective working-class culture.

THE FEMALE LABOR MARKET OF THE 1960S AND 1970S

The export-orientated strategy that President Park Chung Hee presided over in the 1960s and 1970s was, in the words of Hagen Koo, a "capitalist's heaven."[23] State-led development with "the government as entrepreuner"[24] ensured that the state and individual employers shared the goal of building a "rich and strong nation"[25] that required hundreds of thousands of workers. As we shall see, many employers suppressed worker protest with the assistance of the state's security agencies. State security forces could be relied upon to show up quickly in the event of a factory disturbance and to bury protests with effective threats of violence and blacklisting.

From the onset of export-oriented strategies in the early 1960s, the number of women employed in the light-manufacturing sector increased from 182,000 in 1963 to 1.4 million in 1985.[26] And in the key export industries of apparel, textiles, and electronics, young women in their teens and early twenties made up the vast majority of the workforce.[27] These women's presence had already been felt in the first economic plans of the 1960s; by 1970, textiles was South Korea's leading export industry and more than 70 percent of the industry's employees were female.[28] The light-manufacturing sector had its boom in the 1970s when it was, according to political economist Jung-En Woo, "a bridge between the light and heavy phases of industrialization."[29] While the textile sector was still the largest export industry in 1979, it was rapidly being overshadowed by a bourgeoning heavy industry drive that would transform the South Korean economy, and the composition of its working-class, in the 1980s.

The textile industry was one of the most resilient sectors in the period between the end of Japanese colonial rule and the military takeover of the state in 1961 by General Park Chung Hee. It had been resucitated in the 1950s with the assistance of American aid money and was thus able to resume its role as one of South Korea's most important export industries quite soon after the Korean War. The resilience of the textile industry, one of the biggest employers of females, is particularly noteworthy given the received notion of the 1950s and 1960s in South Korea as a period when women returned to their homes following their contribution to the war economy.[30] The triumph of the domestic ideology of "good wife, wise mother" seems most applicable to this period, when after the routing of the left in the 1940s only conservative patriotic women's organizations survived and, in the early 1960s, when the military junta extracted a promise from women's organizations that they would concentrate on charity work or be banned.[31] Although working-class women almost completely disappeared from public culture and political debate in these years, they continued to labor in the fluctuating South Korean economy.[32]

This was also the period when the South Korean state was perfecting the comparative advantage that would make its economy so competitive. The formula was simple: long hours, low pay, and plenty of young people streaming into the cities looking for work. Many of the young women and teenagers entering factories at this time were the daughters of tenant farmers who took the road to Seoul to try their luck in the factory districts and slums of South Korea's industrializing capital. In their own writings, factory girls recorded the drama of their first sight of Seoul. Over the thirty years from the late 1950s until the late 1980s, the population of Seoul increased enormously. In 1972 it was already a metropolis of six million people. With very little international immigration, the population explosion was almost entirely domestic, draining the countryside. Seoul had swelled to ten million people by 1988.[33]

In Seoul and other industrial areas in Inch'ŏn, Masan, and Pusan, young women found themselves central to a labor market that likely seemed both foreign and strangely reminiscent of the social and cultural practices they had left behind in their villages. In her book on female factory workers in South Korea, *Class Struggle, Family Struggle,* the anthropologist Seung-kyung Kim puts it well: "When young women took factory jobs, they acquired an unprecedented public role outside the household, but the low status and meagre wages attached to these jobs was wholly in accordance with the low status with which young women were conventionally regarded."[34] Despite their central role in the South Korean economy in the 1970s, women workers were not regarded by their bosses and their working-class male colleagues as "real workers."[35] Their low pay, their status as temporary workers expected to leave the factory upon marriage, and the limited avenues for promotion for females in the factories led to women occupying the lowest rungs of the factory hierarchy, below the working-class men who supervised them. In fact, an important cause of the defeat of the female-led labor disputes of the 1970s was their sabotage by male workers collaborating with management and the police. Seung-kyung Kim has argued that the women's labor movement in the 1970s was betrayed by male workers who were "distracted by a formal structure of unions designed to co-opt them," while women workers' very exclusion radicalized them to fight their employers and their all-male unions.[36]

THE COUNTRY AND THE CITY

In the 1970s and early 1980s the vast majority of women working in factories were single, and over 90 percent of them were born in the countryside or the provincial cities.[37] The role of many factory girls as a conduit between the country and the city underlines one of the key features of South Korea's successful late industrialization program and the importance of proletarian females both to rural households and to the export economy. As economist Alice Amsden has shown, the labor of wives and daughters sustained the agricultural economy during South Korea's development drive. When young women left the farms in large numbers, shifting their labor from rural households to commercial enterprises, they helped to preserve the agricultural economy by faithfully remitting their wages back home, while at the same time, as underpaid workers, they made profits for their employers.[38] In effect, female workers subsidized both economies and played an essential stabilizing role throughout the traumatic experience of rapid industrialization.

To the young women, Seoul was the focus of both high hopes and trepidation. They came to the capital at the apex of Park Chung Hee's centralization policies that made Seoul a modern, rich, and powerful city to outshine its rival model city to the north, P'yŏngyang. Their excitement and sense of apprehension is palpable in their writings. However, by the time their stories are in full swing, we see these women's disillusionment with Seoul and all it seems to represent to "the daughters of peasants." Sŏk

Chŏng-nam writes of her life in Seoul, "I gained nothing whatsoever from those several alienating years. Only the shattering of my dreams and the end of my longing for a grand and beautiful Seoul."[39]

Shin Kyŏng-suk writes of being compelled to reassess her family's class position when she makes the journey from the countryside to Seoul. Whereas in the country her family was considered comfortable and respectable, the city teaches her that she is one of the lower classes: "In our village our family was known for its bountiful family rites and its ample meals. We had the largest courtyard in the district and in terms of jars of soy sauce, chickens, bicycles, ducks we were the richest household in the village. But I came to the city and suddenly I was lower class [*hach'ungmin*]. My brother had already been placed inside this contradiction and when I came to the city I entered it too."[40]

In recalling industrializing Seoul of this time, we need to adjust the image evoked by the term "factory." Rather than the gleaming sterile environment of the largest, wealthiest companies, conjure up shacks, warehouses, and hole-in-the-wall businesses divided into multiple hutchlike floors, poorly lit and filled with fluff. Adjust the sense of space suggested by the term "dormitory" and instead picture a tiny room where three or four women sleep side-by-side on the floor; or where one shift of bodies sleeps while another works. In fact, it was not easy for a lone woman in Seoul to gain lodgings if she was not attached to either a school or college, a factory or a brothel. Seoul is revealed by these chroniclers to be a fabled city, where the gates that open onto comfort and success are narrow and the competition in teeming buses and streets is overwhelming. Seoul of the late 1970s and early 1980s emerges in factory girl literature as a place where everyone dreams of distinguishing themselves in a city where the glamour of success is worshipped with a ruthless, self-loathing energy. In this context, these books also become unique confessions of ambition and self-assertion. The question "why writing?," why ambition took literary form, is explored more fully in the next chapter.

In their books, factory girls took upon themselves the task of interpreting the country and the city to each other and to their readers. Their detailed accounts of farm and factory life informed their middle-class readers of the inner lives of working-class women. But their works also established the narrator as the subject of larger social forces in South Korea—the migration experience, militaristic modernization, and late industrialization. The writers detailed their encounter with factory culture and explained why painful working conditions alone were not responsible for the difficulties of their work situation. As people who had grown up in the relative looseness of the countryside's work rituals, they had to accustom themselves to the strange discipline of the shift system:

> On the farm regardless of time, when the day gets dark you go to bed and when dawn breaks you get up, but in this place we become complete slaves of time, regardless of changes in the weather, or if it is dark or light. At midnight or 1:00 A.M., deep in the night, we would be working and go for our meal in the kitchens. For people living on farms this is just unimaginable. Even at the height of a busy working time or eating a meal in the middle of the night, I would sometimes wonder if I had stepped into a strange world and was leading a demented life.[41]

However, political awakening and "class consciousness" do not necessarily begin in the city, these writers warn their readers. I do not mean to suggest that the women had to first come to Seoul and experience capitalism before they could clearly analyze the limitations of their position, as though countless fights over female betterment had not taken place on farms and in childhood. This is particularly the case, Chŏng Mi-suk points out, when it came to girls having to abandon their schooling to make way for a sibling, an experience I consider more fully in the next chapter.[42]

Chang Nam-su gives us a glimpse of how her family and village appear when she comes down from Seoul, a Wonpung factory girl now, to renew her identification papers. Chang has taken the night train and arrives at her village station at eight o'clock in the morning. The platform gate has disappeared due to road construction and she must plod through the dug-open highway to get home. Seoul has altered her sensibilities forever, it seems, and Chang's first glimpse of her old home shocks her: "How can I put into words the return to that shabby house and my destitute family. . . . Our squalid home, so different to what it had once been. But then, perhaps it was because I had seen the grand world of Seoul that our home seemed so wretched by comparison."[43]

Chang Nam-su begins to compare her experiences of poverty in the countryside to poverty in the city. On meeting old friends and neighbors who have remained in the village, she is aghast at their ignorance of the true hardship of the lives of blue-collar workers and the poor in Seoul. In a midnight conversation with her neighbors, she launches into an exposé of Seoul:

> Ha, you folk think that city people live well, but do you know anything about how poor people in the city scrape by? . . . Towering skyscrapers, gorgeous things for sale, glittering department stores, university kids and chauffeured cars, my God its indescribable. But in the midst of all that we the children of peasants go to make our living and our

> life is misery. Picture to yourselves, amid the bright faces of students toting their satchels to school, the sallow faces of workers off to the factories. When students take the bus they have special coupons and the conductresses, even though they are working people themselves, bend over backward to serve them, but we who pay the full fare, do you know how rude they are to us?[44] I hate Seoul.[45]

It is striking that factory girls, who often mentioned their longing for white-collar jobs, do not refer to the service sector, perhaps because many of them had already tried the service industry as bath attendants and waitresses before passing the competitive factory entrance tests. The working conditions of bus girls were especially notorious, and the profession gained a name as the occupation were class divisions were most abrasively enforced.[46] While many factory girls avoided the service sector if they could, they harbored longings for the autonomy that white-collar positions seemed to promise. Low-level clerks or "office flowers" appeared to them to embody all the features absent from factory girls: ornamentation, femininity, and fragility. Every office needed an *agasshi*, or "young lady," as much for the frisson of sexual difference as for fulfilling the menial secretarial tasks of making coffee, answering the telephone, doing errands for the office, and typing correspondence and reports. Chang Nam-su's analysis alerts us to the many layers of snobbery that infected employment society and made commuting, indeed any contact with society at large, potentially humiliating for workers.

The experience of the city—its sights, stimulants, and endless possibilities—provoked Chang Nam-su to compare her life to others around her. When she returns to her natal village we can see her make the painful acknowledgment that her old village is now too small to contain her. This experience of displacement was shared by many country girls in Seoul who had relinquished the rural household as the center of their economic, social, and cultural life for the industrial suburbs and inner-city excitement of Seoul, Inch'ŏn, and Pusan. This displacement that industrial capitalism itself brought about would eventually lead working-class women to create new ways of analyzing their world. From it would come the birth of working-class radicalism and a "labor feminism" that offered new ways of seeing the old society and the beginnings of creating an alternative one.[47]

Just as factory girls were forced to adjust their aspirations once they discovered that the "grand and beautiful Seoul" was not within their reach, so also did many working-class women end up finding their own powerful reasons to feel at home in the capital. In his book *The Country and the City*, Raymond Williams talks about the city as an achievement of the countryside: "Looking up at great buildings that are the centres of power, I find I do not say 'There is your city, your great bourgeois monument, your towering structure of this still precarious civilisation' or I do not only say that: I say also 'This is what men have built, so often magnificently, and is not everything then possible?'"[48] Or expressed less grandiloquently by an activist from the Hai Tai factory union: "Our dream before was to save enough money to build up our home in the village. Now this is all over. My mother and I are now used to city life. Although we are poor, we feel that city life has broadened our outlook and enriched us in many ways."[49]

ENTERING THE FACTORY

To speak of the romantic images of Seoul held by these women is not to say that people came aimlessly or dreamily to the capital, without a careful plan of how and where they might start work. These plans usually included the assistance or patronage of a relative or personal contact, perhaps someone from the same region, or an older girl from the village who had come to Seoul and knew of openings. In the early 1970s there was an oversupply of unskilled labor, and the big companies with the best reputations could afford to pick and choose. In addition, if word got out that a certain company had particularly good conditions or pay, then competition for jobs was fierce.[50] Chŏng Mi-suk makes the important point that a system of recruitment based on personal connections made the employers' position very powerful: "In this sort of situation, where people were employed through personal connections, employees easily became both materially and psychologically in thrall to their employers."[51]

This was particularly the case for young, unmarried girls from the country, schooled in the value of giving respect and obedience to their social superiors. As Chŏng Mi-suk writes, "For these young women it was difficult to see the employer-employee relationship as a contractual one, and not a personal one."[52] In fact, the "personal relationship" shared by employer and employees frequently made management more despotic. However, in the bloated labor market of the 1970s, kin and regional ties were an important means of obtaining casual work or gaining the opportunity to take a factory entrance test like the one described below for Tongil:

> The entry test was stringent. The interview wasn't too difficult but lots of people failed the physical checkup. If you weigh more than 53 kg, or if your height is more than 155 cm you're out. There is a manual test that checks your hand and foot coordination, an eyesight test, hearing tests, tests for color blindness, it went on and on. . . . Of course I can see

> that a physical test is a sensible idea, but the process is just like that of slave dealers in a slave market, buying and selling healthy bodies like merchandise. In the market womenfolk are purchased for being fresh and clean like foodstuffs, and people are in demand according to the amount of physical strength they posses and their ability to do arduous work.[53]

The rigor with which the female form was judged to confirm to strict norms might be contrasted to the actual management of bodies once they were inside the factories. Factory girl literature is sprinkled with descriptions of industrial injuries and how they impaired the beauty of the women afflicted. Rather than a grounds for complaint, injuries were hidden by women so as not to lose the position they had labored for in the hierarchy of skilled and unskilled that determined their place and wages in the factory. Many women also hid their injuries from society at large, although some spoke frankly, with humor, about what had happened to them.[54]

In addition to the recruitment tests stipulated by factories, people also found work by selling their labor at various "human markets," like those that operated outside large railway stations in the Japanese colonial era: "The 'Human Beings Market' is right in the centre of the Peace Market. It is the place where workers like us are sold. At lunch time, buyers gather there to purchase shidas or machine operators.[55] It's not always owners or managers of factories, but sometimes workers themselves exchange jobs with each other there. It is not only the place for job seeking, but also for exchanging information. People with no lunch go there to spend time."[56]

The process of recruitment itself illustrates the sense of position and degree of leeway workers felt they had when they entered the factory floor. It was a process where the employers' unilateral power of selection was taken for granted, and an authoritarian and uneven class relationship was promoted as natural. "How many hours they would work a day, what sort of provisions were available for insurance, or redundancy pay, these and other regulations were never made explicit by companies, and the idea of the women workers themselves demanding this was beyond the imagination, as most would not even have known that such things existed."[57] Chŏng Mi-suk maintains that it was the recruitment process that tacitly ordained the character of employment relations.

Yet the fact remains that within the working-class female labor market factory jobs were a coveted occupation. In the early chapters of *The Road to Seoul,* Song Hyo-sun describes her first jobs in Seoul. She begins as a kitchen hand in a Chinese restaurant, then works behind the counter in a general store, becomes a public baths attendant, and finally makes it to being a factory girl. From this history we have some sense of the variety and hierarchy of jobs available to women. Desire for factory jobs, despite all their horrors, was intense. They were the working-class jobs that kept you firmly within established institutions and farthest away from the uncertainty of the streets.

FEMALE ARCHETYPES

Factory work was only one form of the mobilization of plebeian women into the service of the development state. Women and girls also sought work in the cities as servants, shop girls, secretarial clerks, and, as the service industry in the 1970s embarked on a long boom, as waitresses, bar girls, masseuses, and prostitutes. The public persona of factory girl was established in interaction with the two other female archetypes of this era: the female university student and the prostitute.

In factory girl autobiographies, female university students are painful subjects for factory girls. Chang Nam-su, imprisoned for her union work, writes of a university student who is sharing the same prison wing as her and a collection of girl thieves. The student is awaiting trial for fraud (the guard says: "Beats me how anyone could be taken in by her"), and she has been isolated from the other inmates because of her uncontrollable temper and shrieking. The young woman intrigues and disturbs Chang Nam-su, who is told by the prison guard that she is a university graduate and sometimes breaks into English—the ultimate signifier of class and education in the 1970s. Chang Nam-su's first impulse is to envy the student her accomplishments, but as she learns more about her she changes her mind: "She must have been going mad I think. When she screamed it sounded as though she was being whipped. . . . A chill went through me and I pitied her. Her family when they visited looked so anxious and seemed to flinch away from her."[58] Chang Nam-su begins this section dismayed to find herself in prison and drawn as never before to Christianity to make sense of her predicament. But she ends it with a degree of peace, having been unsettled and changed by some of the characters she meets in prison. The guards, the girl thieves, the university student all give her a sense of a wider world outside of the one she is familiar with. Meeting members of a lumpen proletariat and a representative of the middle class whose life is much more controlled by mental illness than by class status, Chang Nam-su encounters predicaments that her union politics has not prepared her for. She shares this episode of incarceration with her readers and so widens the scope of our understanding of the interests of factory girls.

Sŏk Chŏng-nam writes about her self-consciousness at seeing girls her age, in the summertime, "all dressed up in fashionable clothes and shoes and showing off their beauty. I hated the sight of them as they sauntered down the street laughing with their young men. I wanted to slap them."[59] Part of this extreme sensitivity to the lives of other girls their age was due to the strict enforcement of the distinctions that separated classes. Factory girls, whose attempts to pass themselves off as equal to college students were routinely scorned, greeted with suspicion the female university students who responded to the campus mood of the 1970s and 1980s calling for them to enter factories and become one of the working class, to adopt a working-class femininity. Pak Sun-hui, a Wonpung worker, had this to say about the misunderstandings that plagued the entangling of identities of factory girls and coeds:

> In the 1970s and 1980s there were many university students and intellectuals who sought to emulate factory workers by wearing shabby clothes and work fatigues. But in fact these uniforms connoted something quite different to factory girls. Even though factory girls were apprehensive and embarrassed to wear their uniforms in public, they took very good care of them and sewed and repaired any rends in the cloth. I even heard that many girls when they went to the toilet to relieve themselves would take off their entire uniform so that there was no possiblity of it getting dirty.[60]

While work fatigues were a badge of déclassé solidarity for university students, to factory girls, whose hold on repectability was so tenuous, their neatness and cleanliness was a sign of self-respect.

Song Hyo-sun quotes the lectures that were delivered to factory girls by public intellectuals of the Park Chung Hee era, showing the detail these men went into to establish the correct behavior of young women workers: "Ladies, please avoid wearing blue jeans. When you wear blue jeans you confirm all the bad things people say about factory girls. You look like girls with no family."[61] Kim Won suggests that factory girls' later anti-intellectualism and critique of students stems from having to listen to these pompous and wounding speeches that undermined the very notion of a working-class femininity.[62]

The industrial campaign to bring young women into the factories gave women credit for their labor that helped build a rich and strong nation, but their actual remuneration was beggarly low. With no prospect of meaningful promotion in the garment or textile factories, with no prospect of ever leaving poverty behind, it should come as no surprise that many young women were drawn to sex work, even temporarily. In the absence of available vents for their ambition, or possibilities for making money, thousands of women entered brothels and bars in the booming "entertainment" districts. A publication by industrial missionaries describes this process with characteristic rectitude: "Ten years ago, when there were many houses of ill-fame in Chong Kye-Chon and Chang Shin Dong, a lot of young girls who came from the countryside and had trouble making a living drifted here. These people entered factories and worked as shidas and then as operators. But their wages were terribly low and they had no place to live. They slept leaning against the wall in the hallway near the entrance to the factory. Finally almost all of these girls had to go to the amusement section of town."[63]

As mentioned in chapter 1, economic need and the taint of desperation formed part of the image of working-class women from the early years of colonial industrialization. In his book about his experience in the Samwon textile factory, Yu Dong-wu noted that women continued to be sexually stigmatized by factory work. "There is not a single virgin in the industrial districts" was one of the rumors flying around the industrial areas where women workers lived.[64] The received image of factory girls as fallen women was not unrelated to some factory girls moving into prostitution, but it also expressed the taint of any work, any reason why a woman should be out at night laboring in factories or brothels. In the larger culture, many working-class women were seen to be already prostituted by factory work and the pejorative term *kongsuni,* roughly translated as "factory girl" or "working girl," captures this. A *kongsuni* is a girl who is easily approachable, unprotected by social laws of etiquette and strict rules of honorific address that ostensibly apply to all people who are strangers to each other.[65]

While men frequently drew parallels between working-class women and prostitutes, factory girls rarely did. One exception is Kim Kyŏng-suk, who wrote about prostitutes and factory girls: "You and I are from the same lot, all thrown out by this society. But is it right to live like this without making any protest against the world which treats us like worms?"[66] Women employed in the section of the sex industry that catered to foreigners—either American soldiers or Japanese male tourists—were also sometimes referred to as "industrial soldiers." Like factory girls, they were applauded for laboring "behind the scenes" to raise Korea's GDP. By 1989 as many as one in four working women were estimated to be employed in the South Korean sex industry.[67] Indeed, the commercialization of female prostitution was as much a part of South Korea's economic boom as the labor of women in the top-performing export industries of the 1970s and

1980s. Elaine Kim reports that during the 1970s the Korean government encouraged and publicly praised the women who were employed in *kisaeng* tourism for bringing in precious foreign currency from Japanese and other foreign clients.[68]

In fact, the ubiquity of prostitution in the 1970s and 1980s coexisted with frequent bouts of media hysteria about *all* women—maids, university students, airline stewardesses, actresses, factory girls—moonlighting as prostitutes. All occupations open to women, it seemed, carried the potential of whoring. This rhetoric neatly dovetailed with a related discourse on the dangers of factory work, and Seoul, to lone and naïve country girls. Rather than being contradictory, these discourses were consistent with an industrializing society that could not bear to equate wage earning with independence. In wider society, women engaged in factory work could be referred to codedly as whores or sympathized with as children. Patricia Johnson has written that in Victorian England the figure of the prostitute "haunted" working-class women wherever they went.[69] In a society where females were believed to be "corrupted" by factory work, or any sort of work that exposed them to the dangerous and immoral streets, sexual harrassment greeted them everywhere. As in industrializing England, in South Korea women who refused the cover of patriarchal protection, or were too poor to benefit from it, were huntable creatures. But deep contradictions were at work here. The development myth of "Confucian capitalism" was that factory girls could be protected by the company, living *en famille* in the dormitories, while profit was extracted from them. But when working-class women appeared in danger of asserting their rights as workers, they were sexually intimidated and assualted by thugs hired by the factory, by male coworkers, and by the police.[70]

Why is it that when factory girls entered into political activity, even when it is narrowly defined—as in strikes for back wages, for reinstatement, or for the enforcement of the labor law—they faced violence that often turned sexual? In South Korea the ways in which women laboring in factories and women laboring in brothels were defined—sometimes in opposition to each other, sometimes equated—reveals the instability of the image of a woman worker and the threat she posed to a society that found the prospect of her economic and sexual independence disturbing. The social proximity of blue-collar women and prostitutes is more fully considered in the next chapter, when we examine how the stigma of labor played out in the women's own romantic and sexual relationships, leading them to construct the notion of a factory girl virtue.

CONFUCIAN CAPITALISM

The paternalisic management style of both large and small scale factories was nothing new to girls from conservative patriarchal families. In a structuralist analysis of the forces that subjugate women, Lee Hyo-chae points out how President Park Chung Hee extolled those authoritarian and chauvinist traditions in the family system that offered up females for national use: "Park's educational and cultural policies emphasised the traditional ideology of loyalty and filial piety, granting official awards to self-sacrificing women, faithful daughters-in-law and virtuous wives. The creation of such a cultural milieu justified the social conditions under which the state mobilised women as a cheap labour force, which sacrificed itself for national industrialisation."[71]

The entrepreneur in South Korea frequently adopted the sobriquet "Father" to express his benevolent paternal feelings for his "daughters." The equation of boss with Father and the practice of referring to women workers as one's children reveals gender and class ideologies. By classifying women as children, male bosses and managers implied that the women who worked for them were not independent, politically or economically: "Patriarchal ideology played a particularly important role in socialising women into the type of labour force most desired in the low-wage, labour-intensive export industries—one that was docile, submissive, diligent, persevering, and oblivious to workers' citizenship rights. Thus, rather than being an obstacle, the traditional family system functioned as a crucial mechanism through which a desirable labour force was produced and reproduced for the export industries."[72] Whether in the family or the factory, women were to be represented and "protected" by men and had no need to try to represent themselves.[73]

Paradoxically, the story of the development of South Korea's patriarchal capitalism, where factory owners posed as Father and encouraged their employees to labor for them as though they shared the same blood,[74] is echoed in countless mournful songs and poetry about the actual estrangement of family members.[75] The breakup of families so that juveniles or parents could search for work in the industrial centres was a central feature of paternalistic capitalism in South Korea.

The forced estrangement of families and siblings, when combined with the enforced paternalism of factory culture and its collective and supervised leisure—the television room and library at Tongil, the dormitory surveillance system, the group outings—makes one wonder if the single, independent female was potentially a dangerous figure for her employers. Yet working-class women themselves, and those involved with them, were by no means immune to the appeal of a paternalistic political culture. Even the veteran worker-priest Cho Wha Soon, who labored alongside the Tongil employees and was a main instigator of the campaign for a democratic union, betrays in her language how deep the ideologies were. When she says, "My relationship with the laborers is like that between parent and children," we are in no doubt as to who is the parent and who are the children.[76] Yet later in the same book, Cho elucidates the structures that render women, especially blue-collar women, invisible in political movements, and she calls for a women's movement that will have lower-class women at its center.[77]

For Chang Nam-su the cruelest separation was from her younger sister, Hyŏng-suk. Their household in a tiny rented room in Seoul was broken up when Nam-su entered the Wonpung factory dormitory and Hyŏng-suk traveled to Pusan for a job.[78] Drawn by the promise of education, Hyŏng-suk left Seoul to work in a factory in Pusan that provided schooling for its young workforce. In her autobiography, Nam-su says nothing to dispel her sister's hopes, but she suspects the life of toiling all day and studying each evening will wear out her sister. When Nam-su goes to Hyŏng-suk's school in the mountains to visit her, she is directed to the factory where Hyŏng-suk works. Her fears for her sister are realized: "As soon as [Hyŏng-suk] caught sight of me standing by the gates she cried 'Sister!' and flew to me and threw her arms around my neck. Tears were pouring down my face. 'It's been hard for you, hasn't it?' She didn't answer and my throat couldn't open to let me speak. I only had to look at her poor gnarled hands to know how much she had suffered."[79]

Chang Nam-su's experience of rapid industrialization in her family, both in childhood and early adolescence, is described in terms of the separation of siblings and parents. The form of Chang Nam-su's *Lost Workplace* leads from rural family poverty to the journey to Seoul, to the family's vulnerability in the city's labor market, finally ending in the father's breakdown and defeated return to the provinces. It is only after describing the breakup of her family and her isolation in Seoul's employment society that the author is able to fully account for her relief at discovering the labor union. The relief Chang Nam-su finds in the union—the solidarity of the factory floor and the fellowship of friendship and collective dreams—becomes the body of her book, the answer to grief and hardship suffered in solitude.

Factories, on the outside at least, formed part of the iconography of South Korea's drive for prosperity.[80] Their planned space, even down to the body-heated dormitories, were testament to the efficiency principles shared by factory owners and economic planners who believed they could harness their employees' very rest and ablutions to the time-discipline rule of modern capitalism.[81] And the discipline of South Korea's military modernization was indeed immense. In the decade from 1976 to 1985, the stated average workweek for employees in the manufacturing sector was 53.3 hours.[82] And there are many cases of women and men employed in the garment factories working 60 hours a week, and even as much as 80 hours, when large orders needed to be filled quickly. The long workweek was not limited to blue-collar workers either: schools, banks, and government departments all operated on a six-day week, and Saturday half-day attendance was mandatory for both employees and school pupils.[83]

The working hours of males employed in the manufacturing sector were similar to what females worked, but the remuneration men received was more than twice that of factory girls. In 1980 women in the light-manufacturing industries earned 44.5 percent of male earnings, and for most of the 1970s and 1980s they received less than half what male workers received.[84] Conditions in the garment factories were particularly bad. It was in the apparel factories in the Peace Market near Seoul's East Gate that working people's anger first exploded into the open, in 1970. In that year a young garment worker called Chŏn T'ae-il took his own life during a street demonstration in a desperate effort to bring public attention to the conditions in the garment factories. Suicide was Chŏn T'ae-il's final declaration in the campaign he and his work colleagues were waging to force implementation of existing labor law.

The shock at the manner of his death—Chŏn T'ae-il burned himself alive while clutching a copy of the Labor Standards Law—continues to haunt South Korea today. Every Labor Day his diary is read over the radio, and his name and memory reverberate in almost every labor demonstration. Chŏn T'ae-il was twelve when he left school to become a paper seller and a shoe-shine boy at Toksu Palace, but he wrote prodigiously.[85] The mix of threadbare jobs was common practice for male child workers who could not support themselves by one occupation alone. Like Chŏn T'ae-il, they might shine shoes during the day, sell the evening paper to commuters going home, and at night collect and sell cigarette butts from Myondong's alleyways and outside the Midopa Department Store and the Chosŏn Hotel. Chŏn T'ae-il wrote a diary and drafts of several novels. He conducted a comprehensive questionnaire on the lives of his colleagues in the Peace Market, and he also wrote poetry. Revealed in his writings as a complex and principled person, Chŏn T'ae-il became a lasting symbol of the youth who, in the 1960s and 1970s, came to Seoul to seek their livelihoods in factories and on the streets and experienced brutal exploitation.

In his diary, which was circulated after his death and later published, Chŏn T'ae-il reached out to young people in higher classes. He wrote of his longing for a student friend his own age with whom he could discuss ideas and study the abstruse labor law. Almost immediately after his death, his wish was answered for his fellow workers even if not for himself: students who had heard rumors of the demonstration came to the hospital where he was lying and to his funeral in Peony Garden. It was to be the beginning of a new relationship between working-class and middle-class youth, the ramifications of which I explore more fully in the next chapter.

In the 1970s and 1980s, conditions in the garment factories remained bad: dust from threads and the absence of windows, proper light, and ventilation meant that workers frequently suffered from a variety of health problems, including gastointestinal problems, eye disorders, and tuberculosis. Industrial accidents were a major worry in all blue-collar industries, where an abusive work culture and lack of even minimal safety precautions meant that one could lose a finger, or an eye, to a machine or a manager, or even to a coworker. For example, Seung-kyung Kim reports how one manager sometimes threw scissors at workers, and Cho Wha Soon writes of the terrible fights between bus girls at an Inch'ŏn bus company.[86]

Many commentators have noted the militarized management style of industrial enterprises in South Korea during the years of the military-presidents (1961–92).[87] Both Seung-kyung Kim and Hagen Koo have also remarked on the presence of precapitalist practices in South Korean industries, in particular a patriarchal authoritarianism where "workers were looked on not as the sellers of their labor with their own contractural rights, but as children or as traditional servants."[88] Yet the reciprocity of the ideal traditional patriarchal system, where the patriarch used his power to protect those who labored for him, was missing in the factories. Instead, employers preached a Korean-style management that saw the company adopt its young workforce as "dutiful daughters," house them and feed them and bid them work till their bladders leaked, and till their lungs filled with fluff. When working-class women began to strike, they ripped open the contradictions in a society whose traditions were based on a belief in the protection of women but where the factory system exploited them ruthlessly.[89]

It was at these moments, when factory girls asserted themselves as workers, that they exposed the contradictions in the class and gender ideologies of "Confucian capitalism."[90] The originality of Confucian capitalism was that it thrived by grafting Korea's feudal hierarchies onto the modern factory system, giving the whole industrialization process a veneer of cultural continuity. For factory girls this continuity was their economic and social marginalization.[91] The pattern of gender hierarchy in the light-manufacturing industries was defined in ways that provided continuity with older patriarchal hierarchies, with women workers at the bottom.

Seung-kyung Kim notes that the gender hierarchy that operated in the electronics factories in Masan in the late 1980s delegated unskilled work to women while ensuring that their supervisors were almost always men.[92] This notion of "women's work," the idea that the labor of men and women is naturally divided by their gender, permeated all segments of South Korea's segregated employment society—down to single-sex schools and people's hobbies.[93] Kim explains: "The subordination of women within the workplace seemed natural or common sense because it derived from the traditional hierarchical relationship between the genders that permeated society outside the workplace."[94]

But the gender hierarchy was not the only hierarchy operating in the factories. Sŏk Chŏng-nam writes about the torments of the pecking order; and Seung-kyung Kim, in her ethnography, records regional discrimination as well as status distinctions among women on the basis of education levels, whether one had a fiancé or not, how eligible one's boyfriend was, and so on. There were other areas of gender discrimination aside from wage disparity, lack of promotion, and the menial nature of jobs for females. Young women were also considered temporary workers, whose true vocation was domestic. If poverty forced them to seek waged work, then ultimately even that work led them back into the family, whether through saving for their dowry, or supporting siblings who went to school, or paying off family debts.

Cho Wha Soon wrote candidly of her first experience in a factory—in the kitchens of the Tongil Textile Company. Cho was a university-educated Uniting church minister who entered the factory as an industrial chaplain to "develop a theology of mission to the workers," and this required laboring in the factory for six months, alongside other workers. Here, she has just been moved from the kitchens to the textile-making department, the elite section of the factory, and she endures teasing and scolding from workers much younger than herself:

> Even in that kind of situation I was thinking that I should evangelize this factory. Wasn't the purpose of my coming here for this working-training to lead these workers to God? It wasn't just for the purpose of labor itself. "Young lady, how old are you? Where is your hometown?

> Are your parents living?" With a friendly smile, I tried to talk to my younger workmates, while keeping my own hands busy. I was interrupted immediately by the sound of a whistle from somewhere. Startled, I looked in that direction to see the supervisor pointing his finger at me and shouting, "You, who are here for the first time! Why do you have so much to talk about?" It was my first time to be insulted in front of so many people.[95]

Cho Wha Soon wrote that working at the Tongil factory was a revelation. As a middle-class, educated young woman, she had never experienced the feeling of being stripped of one's dignity that was the initiate's introduction to factory work. "Under such circumstances . . . anyone who acted gentle or refined was probably not normal," Cho states bluntly.[96]

Yet for all the alienation, the loneliness, and opressiveness of their situation, there were consolations for factory girls. Chief among these appears to have been the friendships they formed with each other. The urgent pace of rapid, late industrialization threw together village girls from southwest Chŏlla Province and southeast Kyŏngsang Province, and factory dormitories like the one at Tongil gathered three hundred young women to eat, sleep, work, and holiday together. Chŏng Mi-suk argues that alongside the anonymity and alienation of Seoul and the factory districts were the sustaining relations with work colleagues.[97] In factories like Tongil, Bando, Wonpung, and YH, the factory girls' loyalty to each other played a key role in their campaigns to build a democratic union in their workplaces.

LANGUAGE OF DISSENT

Many analysts of the South Korean women's labor unrest in the 1970s have commented on the disconnected, spontaneous nature of their disputes, given a retrospective continuity by the labor movement that would later claim them as their own. In fact, as Yi Ok-ji notes, the disputes were autonomous and spontaneous precisely because no one at the time would back them: "The upper strata of the organized labor movement took the side of the government and the industrialists in obstructing the autonomous [largely women-led] union movement coming out of the factories and intervened directly to destroy the campaign to reform labor conditions."[98]

It is for these reasons that the organizational support the workers received from Christian groups proved crucial. In the years after the Korean War, South Korean society was permeated by a profound hostility and fear of anything that smacked of "communism," a fear that was reproduced and disseminated by some of the most powerful institutions in society—

the state, the army, schools, religious organizations, and the leading daily newspapers. Pak Myŏng-rim, a historian of the war and anticommunism in South Korea, describes how the conflict "accomplished the foundation of a strong anticommunist dictatorship in the South where the ruling government's ideologies of anticommunism and anti–North Korea became impervious to dispute."[99] Yet the tradition of radicalism in the first half of the century did not disappear completely in South Korea. The specter of communism and the terrible memory of the war with the North were kept alive by the state and haunted Seoul in anticommunist banners slung over overpasses and government buildings. For the conservative government of Syngman Rhee (first president of South Korea, 1948–60), and for the military presidents who followed, communism was an enemy to continually be vanquished.

In this atmosphere the language of class conflict was not part of public parlance.[100] When the first "industrial missionaries" arrived in the factories in the early 1960s, they came with a very different mission than what the socialists had brought in the 1920s. Where socialists had seen the working class as the people upon whom capitalist industrialization depends for its existence, and as making up the most powerful potential force in a capitalist society, worker-priests were drawn to laborers because they saw them as the most downtrodden and suffering members of society, as the people most in need of Christian care. But this attitude would change. In her extraordinary tale, worker-priest Cho Wha Soon details how she moved from a superior do-gooding "church-centered tendency" to become part of the democratic revelation in the women's factories. Together with the workers with whom she ran night classes, Cho discovered that self-assertion was the beginning of democratic practice. "We must get rid of such titles as 'intellectual' and 'minister,'" she said. "The true way is to learn through the laborers and make the movement with them."[101] The force ranged against women workers when they tried to create a union at the Tongil factory led Cho Wha Soon to acknowledge the strength and vital role of the factory girls: "nothing causes more apprehension in a capitalist society than the labourers," she concluded.[102] Yet the question remains, with the language of class conflict impermissible, how did these industrial missionaries, and the workers themselves, frame their campaigns? How did they speak politics? Let us turn to examine those disputes.

Industrial disputes by women workers in the 1970s took place in the largest and most prestigious factories. The YH Trading Company was the biggest wig manufacturer in South Korea in 1970, with 4,000 employees, although by the time of its attempted closure in 1979 it had shrunk to around 1,000. The Tongil Textile Company was first established in the colonial era and at the time of the disputes was one of the largest textile companies in the country, employing around 1,300 people, over 1,000 of whom were women.[103] The Bando Garment Company in Inch'ŏn employed 800 workers in 1977, and Hai Tai Confectionery had 2,500 employees in 1976 at the time of the protests over working conditions.[104]

The Tongil textile workers' attempts to elect their own female union representatives provoked a vicious response from the factory's management, and the story of this union is the one most frequently singled out as emblematic of the 1970s labor movement.[105] The Tongil dispute is where the ugly nature of Korean-style "Confucian capitalism" showed itself. In this dispute, young working-class women exposed the misogynist ideologies that the state, capitalists, and working-class men labored together to uphold and make so readily acceptable that they became the invisible "common sense" of a community. Yet the women rose up in such a way that called down all the violence upon themselves. How did this come about?

In 1972 the first female chairperson of a union in South Korea was elected at the Tongil Textile Company.[106] It was the culmination of three years of labor education in workers' rights, facilitated by Cho Wha Soon, which had greatly increased the confidence of the factory girls. When the young women learned that a union could represent them and through the solidarity of a union they could demand changes to their conditions and their pay, they turned their attention to the union elections. The incumbent union was an unobtrusive organization, and its leadership positions were filled by men handpicked by the company's managers as reliable negotiators, men who did not regard factory girls as their constituency.

Knowing this, the women secretly organized their own candidates, and on the day of the ballot they elected twenty-nine young women to the forty-one delegate positions. Following this election, the delegates went on to vote in a woman chairperson, Chu Kil-ja, and an all-women executive committee. These women represented the most subordinate workers in the factory, and the union now acted for them rather than for male line leaders and supervisors.[107] Following this, Korean Central Intelligence Agency (KCIA) men began to trail the delegates and Cho Wha Soon. After the election the delegates and the women workers generally suffered verbal abuse, harassment, intimidation, and the threat of dismissal, while their positions in the factory were continually shuffled around. Their main achievement was to survive and campaign again in the next round, and in 1975 they elected another woman as chairperson, Yi Yŏng-suk. At the election of delegates, however, trouble awaited them.

The election of delegates was in July 1976, but male workers who had been bribed by the company leveled charges against Yi Yŏng-suk for instigating workers to strike, and then they hastily convened the election in the absence of the women delegates and their supporters, who had been locked into the women's dormitory at the Tongil factory. When they heard a company man had won in their absence, the women broke down the doors of the dormitory, forced their way into the factory, and began a sit-in. The next day, July 24, 1976, they were joined by hundreds more—eight hundred in the sit-in strike inside the factory and three hundred at the gates—while the company shut off all water and electricity and locked the toilets. On the third day, traffic was diverted from the area, and families trying to pass in water and menstrual pads to the women were pushed away by the police. In the evening, a bus of riot police drove up and, in the face of the police clubs and helmets, the women took off their clothes.

The substitution of nudity for political demands seems to show that the women shared the same language of female inconsequence as articulated in broader society. Yet this message—"we are helpless, we are defenseless, we are only women"—rather than quieting their adversaries, incited them. The women wanted to shame the police, and those whom the police were defending, but they miscalculated their adversaries. The women were beaten badly, many were sexually assaulted, seventy-two were arrested, and fourteen were hospitalized.[108] Participants, historians, novelists, and activists have returned to this episode again and again to try to make sense of it. Why did the women workers' protest take this particular form? To take off their clothes in protest, in desperation, was in a sense to bring to a crisis the conflicting ideologies surrounding laboring women that were operative in larger South Korean society: one being the noble filial girl, laboring for family and nation while asking for nothing for herself; and the other the female already prostituted by factory work and the slums who could not be degraded by nudity. Ultimately, the women showed that to protest is to disturb layers of interconnecting and resonating ideologies intent on putting lower-class women in their place as "the most degraded members of a degraded class."[109]

The Tongil autonomous union did survive this assualt and go on to elect more women delegates, but its achievements in the workplace were few: it negotiated menstrual leave for female workers; in 1976 it won a breakfast break for early morning shift workers; and later it managed to prohibit physical assault by foremen.[110] But the autonomous union was continually beseiged by the company, by the textile union's bureaucracy, and by the state security forces. Its main achievement, according to Cho Wha

Soon, was that it brought the collaboration of the company, the industry union, and the state out into the open in South Korean society.[111] When Sŏk Chŏng-nam went to the main union body to seek assistance, this was what she found: "When we went to Han'guk Nochŏng [the main union][112] to request cooperation, some office worker listening in asked, 'That's the factory where the women workers took off their clothes and protested in the nude right? When women behave so shamelessly in front of a group of men, well . . . ' They were not interested in knowing why if we hadn't gone that far our protest would have been to no avail, but just stressed the fact that we were women taking off our clothes."[113]

But it was the YH dispute in 1979 that brought working-class women onto the national political stage. Owing to internal mismanagement and embezzlement, the YH Trading Company, once the biggest wig manufacturer in South Korea, announced in would close its doors by the end of April, in the middle of negotiations over pay raises for its employees. Faced with imminent unemployment, the workers and their union, one of the new independent unions formed in 1975, decided to fight. First they took their case to the government on April 4, 1979: "When we visited the Office of Labour Affairs in northern Seoul, we were thrown into desperation upon being told that in a capitalist society, nobody can interfere with the entrepreneur's voluntary closing of his own business."[114]

On April 13 the union called a general meeting, and five hundred workers began a sit-in strike, calling on the company, the bank, and the Labor Office to negotiate the closure of the plant. The T'aenŭng police branch broke up the strike violently, but YH employees continued their strategy, working the day shift and conducting a sit-in stike in the evening. When negotiations broke down in early August, as management announced the factory would close on August 6 and the police were set to return and rout the strikers, the workers played their final card.

At dawn on August 9, about two hundred YH workers shifted their demonstration to the offices of the opposition New Democratic Party (NDP) in Map'o, led by then-opposition member Kim Young Sam (later president of South Korea). When police surrounded the building and Kim Young Sam announced his support for the striking workers, newspapers put the dispute on the front pages.[115] On August 11, one thousand riot police broke into the building and attacked all inside—the YH workers as well as opposition parliamentarians and journalists. In the attack, one worker, Kim Kyŏng-suk, died when she fell from the fourth floor of the NDP building.[116]

The YH workers' involvement of the NDP in their campaign triggered a political crisis. When the NDP took the dispute to parliament and demanded that police involvement in the death of Kim Kyŏng-suk be fully invesitgated, the party's leader, Kim Young Sam, was expelled from the Assembly on October 4. In the highly regionalized political system of South Korea, Kim's constituency in south Kyŏngsang Province came out in the thousands to support him, and President Park Chung Hee imposed martial law in the Pusan-Masan area. Park Chung Hee had built a political system incapable of tolerating dissent and he planned to send in force, paratroopers if necessary, to put down the demonstrations that every day were widening to include anger at the entire Yushin system,[117] at Park himself, and at high unemployment levels. On October 26, 1979, before he could send in the army, Park Chung Hee was assasinated by his own security chief.

Despite the publicity it generated, the YH dispute ended terribly for the workers. Four unionists were imprisoned, one woman was dead, and 223 employees were banished from Seoul. Though the YH dispute was a catalyst for the fall of the Park regime, the workers' concerns were not answered or even addressed by the next regime of Major General Chun Doo Hwan. That the YH workers were the first to involve politicians in a working-class campaign and bring the concerns of blue-collar women into parliament only indicates how isolated party politics was from working-class women in South Korea.

LEADING THE LABOR MOVEMENT

For a few short years in the 1970s, women workers in scattered industries in South Korea stood up for democratic freedom in the workplace and took the full brunt of the Yushin police state's brutality. Most of the other radical, antigovernment organizations in the 1970s were secret, underground cells.[118] Only factory girls faced their employers, and behind them the regime and its secret police, in open defiance. For this they have been lauded by the dissident movements that gained strength from their example—both the democratic union movement and the student movement. What the women workers won for themselves is harder to calculate.

Seung-kyung Kim does attempt to calculate the gains and losses of the 1970s disputes for democratic unions, and her conclusion is that the women's industrial actions of the 1970s produced martyrs but few victories.[119] She criticizes the tendency within the labor movement, a tendency shared by sympathetic chroniclers, to romanticize women's role, an attitude that concentrates on their victimization rather than on the conflicted part they played.[120]

Kim instead raises some important questions about the factory girls' ambivalent response to their position, their "volatile mixture of deference and defiance."[121] For working-class women, in their writings and protests, reproduced key elements of the gender and class ideologies that they were trying to tear down. After they were attacked and beaten, the factory girls of Tongil wrote an open letter to religious leaders calling for their support. The language of the letter is both radical and reactionary. In it one hears both the voice of serfdom and a new subject struggling to be heard: "Please listen to the desperate cry of our poor workers, who are struggling to live like [respectable] human beings despite society's cold treatment. . . . We want to live like human beings, and although we are poor and uneducated, we have learned about justice and democracy through our union. Are we wrong to engage in our desperate struggle in order to keep our conscience alive and not to surrender to injustices? We want to hear your honourable judgement. Please give us your generous encouragement."[122]

On the tactics of the women's labor movement Seung-kyung Kim notes that "women workers also made use of culturally salient images of female vulnerability even as they pursued militant demands in their collective actions."[123] As mentioned earlier, the poverty of the women's political vocabulary was in part a symptom of South Korea's anticommunist state. The tight control over books, classrooms, speeches, newspapers, and even private gatherings impeded working-class women's allies as well. When Cho Wha Soon was arrested for giving a speech on the struggle of the Tongil workers at the Catholic-Protestant Autumn Lecture in Pusan in 1978, her defense at her trial was to deflect political questions into theological ones.[124] In her court trial, Cho used apocalyptic language to make points that politically she couldn't.[125] In a way, theology was her last line of defense against the prosecutors and the KCIA interrogators.

By Cho Wha Soon's own account, theology was often a very powerful aide in her moral and spiritual battle against the oppressors of workers, but it left her and her comrades utterly exposed. Her lack of political analysis meant that while she was empowering her fellow workers to seek a just wage and a democratic union she was also relentlessly exposing them to a police force and its employers that were unmoved by moral or theological considerations. Again and again in the course of the Tongil struggle, Cho Wha Soon trusted in the fairness of government labor committees and even, amazingly, the police. As a campaign where all the risks were taken by the factory girls, and passive resistance tactics like fasting and nudity threw the entire toll of the dispute onto the bodies of the women, the Tongil struggle illustrates the high stakes for the women engaged in this labor union campaign. But women who had been involved in the labor disputes addressed their society through other means too. They turned to literature, and to autobiography in particular, to articulate their lives in a form that gave them some cultural authority. Harnessing the power of representation, they used language to depict their full, ambivalent selves. These authors show how the very elevation of working-class women as suffering symbols (by the labor movement) or dutiful daughters (by the government) imprisoned them in a symbolism that did not allow for the expression of their own conflicted agency. It is time to turn to the women's own response to their social world and the issues that stirred them—the pursuit of education, the consolations of literature, and the sexual politics of poverty.

4. Slum Romance

> I heard a shocking thing today. I had stepped off the bus and was making my way home when I heard "Hey! *Kongsuni* [factory girl]!" A group of male students were loafing nearby.
>
> "Hey *kongsuni!* What are you gazing at? You worthless bitch."
>
> There was no way I would take that so I replied, "So I'm a *kongsuni.* So what have you got to be proud of? Just because you scraped into school does that make you scholars? What a load of crap. You should improve your minds, you losers."
>
> But they kept going with their drivel.
>
> "You ignorant bitch, what do you know? This ignorant bitch, how dare she. . . . "
>
> I swallowed down the abuse I wanted to give them, and went on my way. As soon as I got home I burst into tears I was so outraged. Why do we have to hear such things? Why do we get called this *kongsuni?*
>
> NA PO-SUN, *Uridŭl Kajin kŏt Pirok Chŏgŏdo* (Even though We Don't Have Much)

As the lynchpin of South Korea's first stage of export-oriented industrialization, factory girls were simultaneously thanked and dismissed, elevated and patronized as symbols of a patriotic and selfless devotion to national/family prosperity. In fact, not only did the political suppression of factory girls in all-male unions and their economic marginalization in dead-end jobs obscure the significance of their role in late-industrializing South Korea, but they were central to the rapid development project in so many ways that seemed to efface them—as low-paid, diligent workers; as financial supporters of rural households; as young females slotted in at the bottom of the factory's gender hierarchy reinforcing and reproducing older patriarchal hierarchies. When in the 1970s factory girls asserted themselves as workers, exposing the contradictions in their society that effaced them at the same time as it leaned upon them so heavily, they provoked the wrath of their employers, of the police, and sometimes of fellow male workers. The dissident labor movement responded to their endeavors for democratic relations in the workplace and in wider society by sentimentalizing them and fitting them into the "feminized position of victim."[1] For these reasons, the writings of working-class women themselves that emerged in the 1970s and 1980s are crucial to understanding these women's lives and times in these years. Marginal to the metanarratives of national development and of class struggle, factory girls created their own narratives and in so doing wrote searing commentaries on South Korea's industrialization experience.

This chapter takes as its archive the autobiographical literature of South Korean working-class women to examine the effects of rapid industrialization more intimately. In particular I focus on two recurring dilemmas for these authors: their precarious standing in South Korea's literary world and their struggle to articulate a "working-class femininity." The authors endeavored to construct a literature of agency in a literary and social milieu that barely acknowledged the interior world of lower-class women. They shared intimate truths with a society that was highly exploitative of the provincial girls who came to work in Seoul's factories. Even as they ventured into literature, these authors struggled with the persona of "factory girl"—an identity loaded with contradictory subtext about poverty, sexual availability, and labor movement heroism. I argue that these autobiographies, which at the time were read as straight accounts of working-class female suffering in a repressive state, question deeply held assumptions about culture and the feminine in South Korea's industrializing milieu. Chang Nam-su, Sŏk Chŏng-nam, and Song Hyo-sun employed autobiography to unravel the contradictions they lived in South Korea's industrializing society. In these autobiographies, we see how factory girls embodied all the things that women should and should not be—filial, hardworking, and selfless while at the same time tainted by images of working-class women as corrupt, solitary, and unfeminine. They critiqued their society in tales of hard-won education, longing for cultural literacy, and doomed love affairs with males from higher classes.

These autobiographies were written in a maelstrom of economic, social, and sexual upheaval at a time when there was no clear language to describe nonetheless ubiquitous sexual harassment or sexual assault. Rather than focusing on the widespread sexual violence, or the threat of predatory behavior that overshadowed South Korea's military-run society, these writings instead adopt a defensive tone to expound upon factory girl virtue. The authors wrestle with the paradox of being condemned as unfeminine because they engage in manual labor, while at the same time their economic vulnerability overdetermines their sexuality. It is when we examine the problem of romance in factory girl writings that the sexual politics of poverty come clearly into view,. Namely, how does one navigate industrial life as a sexed subject; or, more simply, what happens to *kongsuni* who fall in love?

BECOMING A WRITER

In Seoul in 1984, an autobiography was published by one of the leading literary publishers in South Korea, titled *Ppaeatkin Ilt'ŏ* (The Lost Workplace). Early in the book the author, a young woman named Chang Nam-su, recounts how she was removed from school and sent out to work. Determined to continue her own course of reading, she describes her thirteen-year-old self, after a day of monotonous labor in the fields, up a tree astride a branch reading *Tess of the d'Urbervilles.* In this one scene of a farm girl's curiosity to see her class depicted in canonical fiction, the connection between literature and self-knowledge is laid before us.

Chang Nam-su wrote her memoir in the hiatus following her release from prison for involvement in a labor dispute at her factory, Wonpung Textiles. Her subsequent blacklisting and removal from productive employment coincided with an enormous upswing of interest in the new genre of labor literature. Even as the cost of critiquing South Korea's version of capitalist development climbed higher, the appetite for true stories from the factory districts increased. Labor literature of the 1980s was the latest in a series of literary movements that had sought to represent the figure of the proletarian woman. At once the lynchpin of a manufacturing-based rapid industrialization economy and a figure of eloquent critique, the factory girl seemed to contain the key to evaluating the contradictory costs and boons of capitalist development in South Korea.

The central role of working-class women in South Korea's rapid industrialization from the 1960s to the 1980s has been obscured by their marginalization in poorly paid, self-effacing jobs. Socially it is hard to find evidence of their value or significance anywhere—they left school early, and their concerns were largely ignored in the mainstream press. Even on their way to work, they tell us, they were taunted and harassed in the streets for being *kongsuni,* degraded because of their need for money and the manual labor they engaged in. It was against this obscurity that working women wrote autobiographies, constructing the self in the context of South Korea's state-led rapid industrialization that asked young blue-collar men and women to sacrifice themselves for a future prosperity.[2] The contradictions that factory girls were caught in are already familiar: these women were central to the export market and the wages they remitted home helped sustain the rural economy, but they were never seen as "real workers," and they held temporary positions at the bottom rungs of the factory. When laboring women began to assert themselves as workers in union campaigns at the Tongil and YH factories and other worksites, the response from factory owners and the police was ferocious. The suppression of working-class women was a key feature of South Korea's successful export-led development, while the opposition strategies the women employed reveal them as both critical of and enmeshed in the gender and class ideologies that were part of their lives.

Autobiography, as the very symbol of self-representation, was itself a stand against the social world that had for so long overlooked them. Yet this literature's counterhegemonic position is more complex than a simple binary opposition to "high culture" and state censorship in the 1980s.[3] For not only was the state under President Chun Doo Hwan (1980–88) itself struggling to attain hegemony in the years these books were published and distributed,[4] but high culture, or at least one stream of literary culture, was turning to worker narratives and *minjung misul*, or "people's art movements," as a source of revitalization.

The autobiographies I examine are representative of the cultural creativity coming from working-class people throughout the 1980s. Plays, street theater, public shaman ceremonies, music troupes, collections of essays, short stories, poetry anthologies and even a movie, *P'aŏp Chŏnya* (The Night before the Strike) were performed or circulated informally as part of a substantial dissident cultural market in South Korea. They were part of a wider people's movement that in the 1970s and 1980s sought to popularize indigenous, "plebeian" cultural traditions and make them part of everyday life.[5] The three books I discuss were all published by "progressive" commercial publishing houses and were sold in the country's largest bookstores—Kyobo Mungo and Chongno Sŏjŏm.[6]

The literary historian Kwŏn Yŏng-min includes labor literature in a wider category of literature spawned by South Korean industrialization:

> As we passed through the 1980s, fictional accounts of the situation of the labor problem and the conventional form [of novels themselves] underwent change. With the beginning of the democratization movement and the opening of the political system, [workers' literature] gave a lifelike reflection of the enormous changes brought about by the liquidation of the structure of authoritarian society in the late 1980s. . . . It is accurate to say that the social problems brought about through the process of industrialization were to make workers' very lives and inequalities a matter of concern in literature.[7]

So this factory girl literature was part of a broader labor literature while at the same time it addressed concerns and desires particular to working-class women.

FACTORY GIRL HEROINES

Chang Nam-su begins her autobiography, *The Lost Workplace*, with a history of her family's circumstances when she was growing up in the countryside in Miryang-gun in southern Kyŏngsang Province. From the very preface, Chang writes commandingly. She judges her family relations as "feudal," and when her father is compelled to make the journey to Seoul one lean spring she says this was "the way the first generation of industrial workers began." Her memories of school are vivid and she writes of her regret at the abrupt end of her school years. Throughout these first few pages, Chang draws on a broad lexicon. She describes farm life like a farmer, analyzes her circumstances like someone who had spent a long time in movement circles, and yet writes simply and directly to her readers. It is when, in the midst of a rural idyll, Chang describes herself sitting up in a tree astride a branch reading *Tess* that the first jolt of autobiography hits the reader. Like Chang, the reader is engrossed in a book with a peasant-girl heroine and cannot help but wonder, will her story also end badly?

This moment is the reader's first glimpse of Chang Nam-su's skillful repudiation of the received image of working-class and peasant girls as ciphers, unaware of their plight as portrayed in literature. *Tess* is Thomas Hardy's novel *Tess of the d'Urbervilles*, a story about the pitiful lives of peasant girls in England and their ruin at the hands of men and religion. Chang's favorite books feature the plucky peasant-girl heroine—Tess in the novel named for her and Katusha in Leo Tolstoy's *Resurrection*.

Tess of the d'Urbervilles is an interesting touchstone in literature's movement into the lives and leisure time of maidservants, peasant daughters, and factory girls. Published in England in 1891, the book divided public opinion when it first appeared, with many readers unable to countenance a novel whose main character was, among other things, an unwed mother. But for many English readers of *Tess*, particularly those of the newer reading public—lower-middle-class and working-class females of the late nineteenth and early twentieth centuries—it was a relief to encounter Tess, "a poor working girl with an interesting character, thoughts and personality," writes Edith Hall in her autobiography of servant life in the 1920s. "This was the first serious novel I had read up to this time in which the heroine had not been of 'gentle birth,'" Hall continues, "and the labouring classes as brainless automatons. This book made me feel human and even when my employers talked to me as though I wasn't there, I felt that I could take it; I knew that I could be a person in my own right."[8]

Likewise in South Korea, working-class women reported an ambivalent relationship to literature. When, in *Kongjang-ŭi Pulbitt* (Factory Lights), Sŏk Chŏng-nam describes being introduced to a male poet, her first acquaintance with someone in the literary world, he laughs at her taste in poetry—Byron and the German poet Heinrich Heine—who represented the tastes of educated readers a half century earlier.[9] Sŏk initially loves the library for workers at the Tongil Textile Company because there she can read poetry uninterrupted for a couple of hours and forget about her cramped life and menial job. She is content to exchange her labor at the factory for the hours of "dreaming" in the library.[10] Yet the literature she eventually embarks on writing is not unconnected with her own lived experience. Rather than reinforce the divisions between "labor" and "literature," between what Raymond Williams calls "the values of literature and the lives of working people,"[11] Sŏk Chŏng-nam in her own literature renders these divisions ambiguous, in stories that are pitched as much to factory girls as they are to poets and other possible readers.

Taking a moment to read more deeply into these women's literary influences, we uncover further grounds for ambivalence. Heinrich Heine, the author of "The Silesian Weavers," one of the great strike poems of the nineteenth century, had this to say about the lower classes as readers of his essays and poetry: "I am seized by an unspeakable sadness when I think of the threat by the victorious proletariat to my verses, which will perish together with the entire old romantic world."[12] Workers for Heine are all very well as workers; in fact, as abject workers they perform an important aesthetic function in highlighting the cruelty of industrializing regimes. But as readers they threaten the privilege that allows him to write (against industrialism). And Thomas Hardy, who was credited with creating a new readership in England of lower-middle-class and lower-class working women, introducing them to high culture via the attractive power of his lower-class literary heroines; Thomas Hardy, who in his novels sympathized with and defended his fictional maidservants and the female companions who were seduced and then abandoned by men of a higher class—nevertheless, Hardy still has these women die terrible deaths (both Tess in *Tess of the d'Urbervilles* and Fanny Robin in *Far from the Madding Crowd*). Hardy's heroines are still punished with the severity his society regarded as appropriate.[13]

Such is the space allowed working women in fiction, even when they are its protagonists and heroines. Such is the literature that factory girls read in order to see their own lives. And such is the literary inspiration that served as a background for Korea's factory girl literature. I highlight these influences, not only because Korean factory girl writers noted their impact, but also to gauge the complex politics of translation between one culture of industrialization and another. The language available to them (and to us) for talking about the figure of the factory girl in Korea is not unrelated to earlier and more distant industrialization projects. And if we trace the Korean writers' diverse influences we come across rich grounds for comparative analysis.

Like *Tess of the d'Urbervilles,* Chang Nam-su's *The Lost Workplace* is also about a peasant girl who is caught up in some of the great themes of the times—the journey to Seoul, entering the factory economy, and joining the union movement's culture of dissent. But Chang writes with the confidence of someone aware that she is "making history." It is no coincidence that those who directly experienced the strikes and the crackdown at the Tongil, Wonpung, and YH factories authored the first labor literature. As these young people were dismissed from their jobs, moved to other areas, changed their names and found temporary employment, and began to write of their experiences, they made visible the social structures that had quashed them for so long.

It is such a voice that Raymond Williams refers to when he writes of the prevalence of the autobiographical form in the industrial literature of Wales. "These [working-class] writers, after all, although very conscious of their class situations, were at the same time, within it, exceptional men, and there are central formal features of the autobiography which correspond to this situation: at once the representative and the exceptional account."[14] The coupling of "representative" and "exceptional," the tension between the individual and collective nature of the working-class autobiography, took its toll on the authors. They were engaged in an ambivalent process to disrupt class lines at the same time that they were memorializing a working-class version of life. Jacques Rancière, in his book *The Nights of Labor,* deconstructs the relationships between workers and intellectuals in nineteenth-century France and calls attention to the disruptions that occur when workers try to escape to another kind of life through writing, only to become the embodiment of their class as proletarian authors. Rancière poses the question, "How is it that our deserters, yearning to break away from the constraints of proletarian life, circuitously and paradoxically forged the image and discourse of worker identity?"[15]

This is also a salient question for South Korean working-class authors of the 1970s and 1980s. Caught between honoring the experience of working-class life and fleeing from it, they attempted to reconcile these two endeavors through the genre of autobiography, situating their self-portraits within the social structures that impinged upon their lives. Thus, even in the most intimate and personal episodes in their books they make explicit the social causes of their experiences: the influence of poverty on sexuality, of peasant backgrounds on city shyness, and of family burdens on the decision to take perilous jobs.[16]

Part of the power and novelty of these autobiographies stemmed from the very suppression of writing by young, impoverished females. These young women had truncated educations and were part of a massive juvenile labor market. The very intimacy of the autobiographical genre gave these authors the space to write revealingly of themselves and their circumstances. Sŏk Chŏng-nam wrote about the costs of autobiography—the stigma of revealing one's occupation and consequently one's poverty and vulnerability—to the public.

In *Factory Lights,* Sŏk Chŏng-nam does not reveal the identity of the "poet" who first read her work and encouraged her to publish her diary. Although she only hints at his full name, she nevertheless gives a revealing sketch of his person and character. When a friend of Sŏk's persuades her to meet a poet from the Christian Academy, a religious organization with strong social justice leanings, who expressed an interest in reading her writings, Sŏk fantasizes about what this poet will be like. She imbues her fantasy of him with all the hackneyed, romantic traits of a poet of convention: "I [had] imagined him above all to be a noble individual with long, curly locks cascading down his shoulders and a brilliant light in his haunted eyes."[17] Instead, she tells us with palpable disappointment, her poet was a chubby chap in his thirties. Sŏk is further dismayed when he proceeds to address her in "low" or "informal" language.

The problem of language and address is worth exploring. The use of honorific address forms in Korean is a defining marker of status. Thus the poet, who is older than Sŏk Chŏng-nam, is a well-educated man and of a higher social status than she, has a wide linguistic choice. He can address her using a variety of honorific forms, or he can drop the honorifics altogether. That he chooses to do the latter is not grammatically incorrect but rather illustrates the social status he enjoys. Sŏk must address him with honorifics attached, despite the fact that he shows no such delicacy toward her. It is a choice that annoys and momentarily silences her. That the (mis) use of honorific and informal address forms was a significant issue for many working-class women can be judged by the frequency with which the issue appears in their writings. A film of the era, *Youngja's Glory Days* (1975), provides a striking example of how working-class women were rendered mute in everyday exchanges. In the scene where the future lovers Youngja (a maid) and Chang-su (a worker in a bathhouse) first meet, Chang-su addresses Youngja using *panmal,* the low or intimate form of address. Acutely embarrassed, Youngja does not reply and instead poses the question, "Do Seoul people not know how to address strangers?" as though to a third person in the room. In this scene, age, class, and gender mediate the conversation, showing how difficult it was for a working-class woman to assert herself when the very language employed predisposed condescension.[18]

In her autobiography, Sŏk Chŏng-nam writes of hesitating to give the poet permission to read her diary, but she cannot explain to him the reasons for her reluctance. She explains them to us: "More than anything else what shamed me was that this diary would reveal how destitute, how plagued by poverty my life was. Of course as I was a factory girl making my own living people would naturally assume that I was poor, just as the poet had. But if he were to discover just how degrading my life up until now had been . . . even thinking about the exposure brought goose bumps to my flesh. If it is published there would be no escaping the humiliation."[19] In this disclosure, Sŏk Chŏng-nam is candid about the personal costs of autobiography and the convictions that urge her to dare publication of her diary as an autobiographical story. In entrusting her book to a literary world that had rarely before done justice to representations of working-class women, Sŏk was risking a particular kind of exposure. Added to that was that her first piece, "Pult'anŭn Nunmul" (Burning Tears), an account of the Tongil dispute, was published while she was still working at Tongil Textiles.

At first, people above Sŏk in the factory were proud that one of their employees was to be a published writer. A female guard sought her out and said, "I heard that you have a talent for writing, keep it up," and stroked her hair.[20] But when the piece appeared, the atmosphere quickly changed to one of cool animosity. Sŏk was called in to meet the labor manager who told her she had created "vile propaganda" against Tongil and had been used by the magazine, which only wanted a shocking story to sell its product. She was warned, but not dismissed. Meanwhile, readers of *Wŏlgan Taehwa* (Monthly Dialogue), the literary magazine that published Sŏk's account, were in 1976 reading one of the first pieces of what would become a revitalized labor literature *(nodong munhak)* movement.

We saw in the previous chapter how alone working-class women found themselves to be when they protested their conditions. Their sole supporters when they first began to form unions in the 1970s and attempt to change their circumstances were the urban industrial mission priests like Rev. Cho Wha Soon, who had worked with them in the factory and night schools. Chang Nam-su, who was employed at Wonpung Textiles, reports hearing about the "Tongil excrement incident"[21] in February 1978, not through newspapers or radio coverage but via rumours in the Wonpung factory. When Chang inquires of her colleagues if the newspapers have carried the story, someone retorts, "When would something like that ever be in a newspaper?" And they are correct, as Chang herself concludes: "So the story of the blood curdling screams of the Tongil workers: 'We cannot live on excrement' started to sweep through [the factory districts]. And even after the Tongil incident had become a major social issue, not one line referring to it ever appeared in the newspapers."[22] The very indifference of the established media in South Korea played a role in inspiring the witness accounts in these autobiographical narratives and helped motivate an upsurge in alternative publishing in the late 1970s and 1980s. Out of the suppression of factory girls' writing was born the autodidact culture of working-class South Koreans and the thriving business of night schools.

These autobiographies and the influence they had on readers tells us much about people's disenchantment with Seoul's capitalist society in the 1970s and about who in society would most powerfully express that discontent. The urgent tone of these books and the immediacy and violence of the situations they describe indicate how absorbed both authors and readers were by this new genre. University students had never found themselves described like this before, nor had visiting journalists been able to capture the odor of factory work as these authors described it. Amid all the rumors and propaganda of the Yushin period[23] and its tumultuous aftermath, in these books the morality of "modernization" found itself openly judged by the morality of honorable dissent.

STUDYING UNDER ADVERSITY

In their efforts to continue their education at night while laboring during the day, an activity they dub *kohak* (literally, "studying under adversity"), these proletarian authors advise their readers how hard won was their learning. Education was one way to distinguish oneself from other young women in the poorer classes in Seoul and to scratch off the taint of manual labor. Kim Seung-kyung observed how factory women in Masan in the late 1980s "strove to maintain the status distinction" between factory work and sex work. Educating oneself was one means of acquiring distance from the more violent occupations available for young women.[24]

All three authors—Chang Nam-su, Sŏk Chŏng-nam, and Song Hyo-sun—relate leaving school prematurely as the moment they discovered the limited horizons of the poor.[25] Chang Nam-su was withdrawn at the end of primary school and sent to work in the fields. She scored the highest marks in her grade, but when the school prize at the end of the year went to the daughter of a wealthy family, Chang records it as a valuable lesson on the ways of the world[26] and her first experience of education as social exclusion: "Sometimes, carrying the mown grass balanced on my head or driving the cows home in rows, I saw my old classmates on the road carrying their satchels. I couldn't bear to meet those girl students in their white blouses. On the days when I caught sight of them I would be too rankled to eat dinner."[27]

In many urban and rural families, the education of male children was of paramount importance. The literature of factory girls contains many, many stories of foregoing school to make way for the aspirations of a brother. Chŏng Mi-suk interviewed working women for her study on the South Korean labor movement in the 1970s, and one factory girl related the following personal history: "When older brother was at high school it was our job to carry the yoke and water buckets that adults lifted. We would carry [the buckets] up a hillside, our hearts thumping with the effort. And when I would see older brother stretch out his belly and jabber away in English, I really experienced hate. Older brother was made much of as the pillar of our family, and we were expected to be submissive to him."[28]

While Chang Nam-su "doggedly read books and studied English," furtively keeping up with the curriculum while at the same time working in the fields, many girls making the journey to Seoul took their aspirations for learning with them to the capital.[29] When in 1973 at the age of fifteen Chang began working in a sweets factory, she was called *ŏnni,* or "older sister," by most of the employees in her section. Work began at eight in the morning but there was no official clock-off time at Rolex Confectionary Company. Because she was a *hakwon* student Chang was able to leave at five o'clock, leaving the resentful children to "wrap lollies in bitterness."[30]

Generally speaking, there were three varieties of night schools in the 1970s: factory schools set up by companies at the behest of the government; commercial *hakwon,* or schools that were either specialty cram schools for high school students (geared toward study for a discrete goal, usually passing a university entrance exam) or were patronized by people already in the workforce who wished to continue or "complete" their education; and schools set up by church groups or by radical students or a combination of the two that taught a variety of subjects ranging from classes in labor law, to lessons on Chinese characters, to lectures on the "rights" of workers.

Factory schools arose in response to a shortage of skilled, technical workers in South Korea when, at the time of the fourth five-year plan (1977–81), the government ordered firms that employed more than three hundred workers to set up on-site vocational schools.[31] As mentioned in the previous chapter, textile factories sidestepped this requirement, which they deemed inappropriate for the textile industry's temporary female employees who lost their jobs when they married and faced few opportunities for promotion. Instead, they established in-plant primary schools and sometimes seconday schools. The result, according to economist Alice Amsden, was that "paternalism came to operate round the clock. Factory girls slept and ate in company-owned dormitories, spent nine and one-half hours on the job, and devoted evenings to study in company-owned night schools"[32]

Most *hakwon* were commercial establishments staffed by trained teachers or university graduates. By contrast the *hakwon* where Chang Nam-su and her sister Hyŏng-suk studied offered free tuition and the staff were university students. Chang Nam-su studied with thirty to forty others, all determined to gain their middle-school certificate in a canvas tent, from 5:30 to 9:30 in the evening. Despite the long hours, the windy tent, and the packed classes, Chang was delighted to be a "student" again.[33]

According to historian Chŏn Sang-suk, unskilled, low-paid, and unlettered girls were at a premium in the development market of the 1970s:"For women already studying under adversity, obtaining a job commensurate with one's education [if one had one] was very difficult, while on the other hand the opportunities were legion for barely educated and low-paid production workers."[34] Longing for education, then, had very little to do with employment aspirations. Factory girls did not take the middle-school graduation exams in order to secure promotion in their work.

The third kind of night school was specifically set up for factory workers. Beginning in the 1970s, university students organized these schools "where the basics of 'reading, writing, and arithmetic' were intermixed with analysis of capitalism and political thought."[35] Although these night schools were clandestine, they slowly grew, and they eventually fed into the *hakch'ul* movement that saw students continue their consciousness raising activities inside the factories.[36]

Religious groups were also part of this worker night-school movement. When the Methodist minister Cho Wha Soon first began industrial mission work in the Tongil Textile Company in Inch'ŏn in 1966, education programs for workers were not greeted with suspicion by the company but were actually welcomed.[37] Indeed the worker-priest George Ogle received government and Inch'ŏn City Council commendations in the 1960s for his work in labor education.[38] As Cho Wha Soon writes, in the 1960s it was often the company management that asked the industrial chaplains to set up education programs for employees: "We cannot imagine such a situation now [1988], but at that time, in the late 1960s, it was possible. That was at the beginning of industrial mission and before the labour issue became a general social issue."[39]

Cho Wha Soon's first lecture program, in 1967–68 had the following themes, chosen by the Tongil workers and attended by over 200 people (of 1,300 employees) before or after their shifts: "meeting the opposite sex (this was the subject of most concern), knitting, cooking, handicrafts, flower arranging, [and] home etiquette."[40] As Cho describes in her autobiography, *Let the Weak Be Strong*, the evolution of the night school at Tongil, from offering classes in domesticity to lessons on the labor law, exemplified the democratic mood of the night-school movement and the connections being made between personal and political knowledge. Cho recorded her own education through the industrial struggles she participated in first at Tongil and later at the Bando Textile Company.

Chŏng Mi-suk writes that the young women's shared experience within patriarchal families of foregoing the chance to be educated, and thus their thirst for learning, would later become a strong motivation for their participation in the education programs that church groups and students, and later democratic labor unions, would offer.[41] Often the very labor of factory girls was an expression of their distance from the promise of education. Those who were sending money home to make a gentleman of a brother, and get him into college, might be said to have contributed their labor to his erudition.

The night-school movement's attempt at "majority democratic education" was the focus of a complex range of objectives. It continued the theme of self-improvement so attractive to factory workers who looked for an identity outside the shop floor, or who aspired to possess the cultural capital that education seemed to promise. A major attraction of the church and student-run education programs was that they offered a comprehensive analysis of the world combined with frank acknowledgement of the shared difficulties and aspirations of the young people drawn to their classes.

Yet the factory workers' relationship to education had its own hard lessons. Hoping to escape from a menial destiny, factory girls wrote about how they lost themselves in books and dreamed of office jobs, as Chang Nam-su describes here:

> At that time [while working in the Wonpung factory] I was going to Hallim Hakwon in Yŏngdŭngp'o, which had several students from Wonpung Textiles. I studied abacus calculations diligently and memorized English words and took the Hakwon's school bus, all the while dreaming of when I would progress to enter a good office. Despite the fact that working and studying was more than I could handle, the prospect of only working in the factory was unbearable. Every day I would tell myself, only study hard and one day you can leave this life behind, and I believed it utterly.[42]

Chang would soon find that in the female labor market her education in fact guaranteed her nothing but an intolerable workload: "there was no damn need for a worker who attended night school," she concluded.[43] Eventually Chang was fired by her boss at Rolex Confectionary for being unable to fully devote herself to the job.

Jacques Rancière defines a worker as a person to whom several lives are owed.[44] This is nowhere more evident than in the spasmodic careers of worker-students.

> I remember when I was in elementary school our family would use electricity sparingly to save every cent. I remember getting up in the night, stealthily so as not to alert anyone, and standing by the five-watt globe, shielding the lightbulb with my book and reading until deep into the night. I never felt any discomfort in my legs. I never felt grief like this. Our teacher's words "learning is priceless" *[hyŏngsŏl chikong]* engraved themselves on my mind and I believed that I too would later find success if only I threw myself into my studies. If you work hard there is nothing you cannot do, I believed. How far and long ago those naïve words seem to me now.[45]

The workers' night schools, with their complex aspirations, in many ways complemented the *hakch'ul* movement that saw university students squander their prospects to enter factories as "disguised workers" and labor for revolution from below. While not entirely relinquishing their role as "teacher," university students from the more prosperous classes came to the factory districts to gain an education in the ways of the laboring classes.

WORKERS AND STUDENTS

Chŏn T'ae-il was the first young writer from the sweatshops whose diary found a large readership. After his suicide in 1970, his writings were circulated informally and eventually published in a collection titled *Nae Chukumŭl Hŏtdoei Malla* (Do Not Waste My Life).[46] Chŏn's writings cover a broad spectrum, from reflection on social injustice to poetry to minute accounts of the complexities of poverty. He wrote candidly of the juvenile labor market and in one harrowing scene describes the events leading up to his temporary abandonment of his younger sister one night in Sŏdaemun.[47] For the thousands of people who would read his autobiography, here was someone who had experienced all the contradictions of a rapidly developing economy and society in his own life span. Even in death, Chŏn was caught in contradictions: he who had spent so much of his life campaigning against the physical suffering of young workers brutally took his own life.

Chŏn's writings, his struggles for the enforcement of labor law, and the manner of his death had an enormous impact on the youth who followed him. Chŏn had put into writing his dreams of a different society, his revolutionary longings for equality and friendship between students and workers. It was this dream that captured the imagination of his generation and the generation that followed him, opening up the political imaginations of young people in Seoul and all over South Korea who read smuggled versions of his texts. In being able to articulate, not only what he was against, but what he was for, Chŏn spoke to the collective desires of young people in South Korea. His vision that crossed classes, that found a broad and eclectic readership up and down the country, was one of the inspirations of the *hakch'ul* movement that in the 1980s brought university students into the factories and among the workers they had read about.

Yet the fact that Chŏn T'ae-il, who himself wrote poetry and drafts of novels and became proficient in the details of labor law, should need to appeal to students is worth examining. To Chŏn and those like him who were without a "good education," students represented what learning and literature were. We need only look at the role that education and literature played in the lives of the working-class people we have been discussing to gauge their symbolic significance. Chang Nam-su and her night-school friend Songja gave up sleep night after night to secure their middle-school certificates. And at the age of nineteen, Chang was able to discuss novels and sociology with a university student she met, even though she left the school system when she was twelve. The sacrifices people made to acquire an education that was not of use in the employment market bears out its symbolic value.

But the significance of education to working-class people, and of the students who represented it, was not merely symbolic. Education offered both cultural capital and the tools to critique society systematically. To those like Chŏn T'ae-il, who perceived the need for a comprehensive response to the crisis that was overwhelming him and so many others employed in the brutal "economic miracle," this desire for a student friend carried particular poignancy. Chŏn perhaps foresaw how powerful students and workers might be if they joined together. And so it would be, by the late 1980s, that the student movement and its ally the labor movement emerged as the two most powerful groups to force change in South Korean society. The student movement, with support from the populace, forced the resignation of President Chun Doo Hwan. Over three weeks in June 1987, the country came to a standstill as students, opposition politicians, Christian groups, workers, and segments of the middle class took to the streets to demand democratic change. The demonstrations were followed by a series of wildcat strikes affecting every major industry in the country, with workers demanding free trade unions, better working conditions, higher wages, and supporting the call for democratic reform. Clearly, it was a powerful combination.

Much has been written about the *hakch'ul,* or student-workers, those university students who gave up their advantages—a hard-won university place and a promising career—to become a "disguised worker" and devote themselves to the labor movement.[48] Student-workers were breaking the law by taking factory jobs, as South Korean labor law stipulated that it was illegal for university students or graduates to work in factories. One account published in 1991 estimates that as many as three thousand students labored in factories by the mid-1980s, and the practice continued into the early 1990s.[49]

If we look at the worker-student alliance *(nohak yondae)* from the perspective of workers, we see that it contained many continuities with the old hierarchies it was trying to disrupt. For it was workers who accepted students into working-class life and taught them about the factory world and the slum districts. The *nohak yondae* was a revelation for students, but it appears to have made much less of an impression on workers. They were not the ones trying on a different life, slipping into university as a "disguised intellectual." In many ways the so-called worker-student alliance reinforced how divided the two classes were.

Class mobility did work the other way, from worker to student, but rarely in a style prescribed by the labor or student movements. One notable example is the "fake female university student," or *katcha yŏdaehaksaeng,* who gained notoreity in the 1980s as a product of South Korea's dizzying social change and the anonymity of urban Seoul, where people could conceal their origins and take on new guises if they possessed the requisite ruthlessness and nerve.[50]

In Yun Jŏng-mo's *Koppi* (The Halter), a best-selling novel of 1988, a fake female university student lives on the border between classes, trying in some company to pass herself off as a flower of Korean youth, a student at one of the elite women's universities.[51] In her story her transformation into naïve, innocuous middle-class girl is contingent on unlearning the knowledge she has picked up in the "romance economy" of the entertainment districts. Class transience here becomes a story not of awakening but of concealment.

At a time when overseas travel was out of reach for all but the very privileged, and information about their own society was censored and restricted, it should not surprise us that people took to traveling between classes. This contact between young people, especially those who went out voluntarily to meet each other, was revelatory. Indeed it is in factory girls' encounters with people of other classes, particularly the university students who "have gone out to meet them and perhaps wish to expropriate their role,"[52] that some of the most revealing portraits of Seoul's class-divided streets can be found. Perhaps the most telling examples of the fragile relationship between young people working in factories and people studying at universities are in their love affairs. Here, young female workers reveal how thin-skinned they are about all the things that define them in the abstract—status, employment, poverty, and lack of education. Throughout the 1970s and early 1980s young, unmarried, curious youth flooded into the capital looking for work. In 1983, 72 percent of women in the manufacturing sector were between eighteen and twenty-four years old.[53] Many of them had come alone to Seoul and sent wages home to their families in the provinces.

CLASS ROMANCE

Some of the most avid readers of the new labor literature were university students, who found themselves portrayed in this literature as markers of the divisions between people from different classes. Nowhere were these divisions depicted more forcefully than in the intimate tales of class romance. Episodes of romance in working-class literature are pervaded by a sense of the dangers of romance for factory girls, shadowed by the "romance economy" of the streets. The proximity of red-light districts, the financial and glamorous appeal of employment there, and the many attractive guises of the pimp all added up to make falling in love a potentially perilous business. Working-class women, and all women close to poverty, were haunted by prostitution, part of the shadow economy of South Korea's economic boom and the hunting ground for the country's oversized military class. The extensive prostitution industry spilled over into working-class districts as well as into the middle-class suburbs of Seoul. Shop windows displaying women could be found tucked beneath overpasses and were prominent in university districts as well as nearby all the large railway stations. "The Love Story of Unhŭi," a cautionary tale of class romance from Chang Nam-su's *The Lost Workplace* exemplifies the hazards of falling in love outside of one's class.

Unhŭi is a country girl living and working in Seoul. She is from Chŏlla Province in southwestern South Korea and works at a factory that makes dolls' clothes. On a rare day off, Unhŭi goes with friends to visit the palaces of Seoul she has always dreamed of seeing, and at the museum at Kyŏngbok Palace she meets the affable, attractive university student Sŏngho. They strike up a friendship that soon becomes intense, and despite her friends' warnings Unhŭi devotes herself to Sŏngho—tidying his lodgings and even lending him money for his school fees and expenses. One evening after her shift Unhŭi is waylaid by Sŏngho, who begs her for a favor—can she lend him the 200,000 won he needs urgently for a purpose he asks to remain secret? Unhŭi hesitates, then says yes, without knowing how she will get such a large sum of money, and that night they spend the night together in a hotel and sexually consummate their relationship.

The next morning Unhŭi rises early and goes to the lodgings of her friend Mihwa, who has saved 300,000 won to send to her parents to pay off the family's debts. But Mihwa is not home and the landlady lets Unhŭi inside to wait. In her great haste Unhŭi yields to temptation and takes the money, thinking to explain all to Mihwa later. Unhŭi goes straight to Sŏngho's lodgings but this time the spell between them is broken. There is a new coldness between them, and after giving Sŏngho the money Unhŭi is leaving when she sees another woman—well-dressed and beautifully made up—walk up Sŏngho's lane. This, it seems, is Sŏngho's real girlfriend—a woman of his own class. Sŏngho cannot hide his discomfiture and Unhŭi, overcome with distress, dashes away and reaches home only to find a plainclothes policeman waiting for her. Unhŭi and Chang Nam-su are in prison together when Unhŭi tells her story.

Unhŭi's story of infatuation, theft, and incarceration displays the principle that someone must pay the price for the privilege exercised by a higher class to loaf through an industrial revolution.[54] Factory women like Unhŭi could find themselves bankrupted by this romance economy that paid them for their labor while charging them exorbitantly for their credulity, boredom, loneliness, and consequent craving for leisure and romance. Strictly speaking, Unhŭi is not in the employ of Sŏngho, and she hands money over to him because he asks for it and not because she has earned it for him. But their economic and erotic transaction that exposes her while it enriches him is close enough to the conventional relationship of prostitute and pimp to serve as a warning to all of Chang Nam-su's working-class female readers.

In a nuanced account of her own class romance, Chang Nam-su explores the possibilities of an equal relationship. The historian Chŏng Hyŏn-baek was the first to point out what an intricate picture we can form of South Korean society in the 1970s by following the course of the romance between Chang Nam-su and the university student Hyŏn-u, "a relationship formed through shared interests and wide reading."[55] A love affair culminating in marriage into a higher class was one of the few ways that factory girls could escape the dormitories or tiny slum lodgings and the private degradations of poverty.

The year is 1977 and the author, Chang Nam-su, is nineteen and has just started work at the prestigious Wonpung textile factory in Seoul.[56] Chang has gone home to southern Kyŏngsang Province in the southeastern part of the country to renew her identity papers. She has just said good-bye to her grandmother on the station platform and has boarded the local train for a slow ride back to Seoul, when she is drawn into a conversation:

> The young man in the next seat addressed me.
> "Miss, where are you going?"
> "I go as far as Yŏngdŭngp'o, and where are you going?"[57]
> "That's lucky, I'm going to Yongsan. Shall we talk a little?"
> I smiled as I nodded my head. He was a second-year student at K University and his name was Hyŏn-u. I told him I was working but I also wanted to study more and for the moment was studying [part-time] at a night school.[58]

The night school Chang Nam-su was attending at this time, Hallim Hakwon, was not part of the political night-school movement—a joint venture between radical students and workers—but an ordinary evening class where factory workers could study for their middle-school certificate. Yet this exchange illustrates how education was a flash point for factory girls who had left school unwillingly to earn money or make way for a sibling. To labor by day and study by night, as Chang Nam-su was doing, was to cram two lives into one and put an almost intolerable strain upon one's body. While she never reveals how hard-won is her learning, the flow of knowledge between "worker" and "intellectual" is shown as very much two-way in this exchange:

> "Let's be friends, and lower our language"
> Even though I assented to this I could change my language only after he pointed out to me again and again [that I was using high form]. As we talked I felt that our ideas were similar and I chattered away without resting. We discussed Tolstoy's *Resurrection* and *Demian* by Herman Hesse.[59]

The novels discussed by Nam-su and Hyŏn-u are revealing. *Resurrection,* by Leo Tolstoy, is the story of the spiritual redemption of the dissolute Prince Nekhlyudov who forgoes his early promise when he seduces and then abandons his family's treasured servant, Katusha. Besides being a tale of spiritual atonement, *Resurrection* is also a class love story. Herman Hesse's *Demian* also relates a spiritual journey, where the impressionable author, a schoolboy, meets the enlightened and charismatic Demian at school and falls under the influence of his doctrine: that organized religion and public morality are cant.[60] We ascertain that Hyŏn-u and Nam-su have read both books, as Hyŏn-u knows not to offer himself as the rake Prince Nekhlyudov but rather proposes to play-act the magnetic, Svengali-like Demian:

> "Nam-su, was Demian fantastic?"
> "Yeah, just great."
> "Really? Then, can't I be Demian for you?"
> "Psh . . . Hyŏn-u, no way."
> "Why? Why 'no way'?"
> I was bursting with laughter.
> He was laughing too. I showed him *Monthly Dialogue*[61] and told him to read the article in it called "Human Market."
> "What's it about? Nam-su you tell me."
> So I went into a full explanation for him. Before we knew it the train was passing Anyang.
> "Look, that looks like a business; why does it have its lights on at night?"
> "Because they're working," I retorted brusquely.
> "What! They're working at night? They really keep working at night too?"
> I was struck dumb. How can this person not know that? What kind of a person is he? This "friend" asking if people work into the night? A great gust of loneliness squeezed my heart and made me miserable. Ah, how can this person be so content knowing so little about the world? Are all university students like this?
> He saw my expression. "I really didn't know. That people keep working into the night, I mean. It must be because I have passed life so igno-

rantly. But poverty also has its happiness, doesn't it? I'm convinced that the poor are happier than the rich."

"Ah, so Hyŏn-u took the bumpkin train so he could taste a little of that 'poverty'? What a treat! Do you know a little about real poverty?"

At my sharp retort he bowed his head.

He didn't go to Yongsan but got off with me at Yŏngdŭngp'o.

"We can't part like this. Shall we have something to eat?"

"No thanks, I'm off."

"Well then give me your address. I'll send you a letter."

"No. See you Hyŏn-u."

As I turned away he put out his arms like a child and blocked my way, he wouldn't let me pass.

"OK, then Hyŏn-u, you write down your address for me. I'll send you a letter. That's all right isn't it?"

He had no choice but to take out some paper and write down his address.

"Aiyu, how can your handwriting be this bad?"

But he didn't answer me and just kept writing.

"I shall write," I promised myself.

I waved to him from the bus but he just stood there looking after me. In that empty dawn bus I said to myself, "All right, let's write. And I'll tell you a little more about the world," and soon after that I did write. His reply came immediately.[62]

Nam-su and her new friend Hyŏn-u deepen their acquaintance through correspondence, and make a plan to meet:

> I picked a day and a place that suited me and sent him the letter. On the day of our meeting I arrived at the rendezvous place exactly on time, but he didn't come. I waited for ages and only when my pride was thoroughly bruised did I get on a bus, alighting at the terminal, Suyuri. The April Uprising memorial is nearby. He had talked about the April Uprising. . . . The memorial's white pillars drew my eye. I walked through the graveyard then sat on the lawn and passed several hours just thinking. He had stood me up. I could not bear that on our very first date he had stood me up.[63]

The April Uprising memorial in northern Seoul that Chang Nam-su wanders into is the memorial to students who died in the April Revolution of 1960. Fired on by troops who were ordered to do so by President Syngman Rhee as the students protested election rigging, political corruption, and police violence, the demonstrators soon gained control of the streets and within days brought down the government. The April Uprising memorial has been an important symbol for students of their painful and exemplary past in leading the attack against corrupt regimes. However, this memorial holds little consolation for Chang Nam-su, whose own political acuteness seems so much greater than the student Hyŏn-u's:

> I had been dreaming of him as a possible boyfriend. What a fool I was. He is a university student. He's not someone who has the time to sit and listen to a *kongsuni*[64] like me musing about the world. . . . Two days later a letter arrived. "I am truly sorry for not keeping our appointment on Saturday. Conscript duties prevented me. I had no opportunity even to contact you. . . . "
>
> I knew as soon as I read those words that he couldn't have helped what happened. . . . His voice asking me "Do people work at night?" swirled around in my head. I am a worker in a textile factory and he is a university student. . . . How much I had learned in the space of a few days.
>
> If he had not been a university student I cannot say if my pride would have been so wounded, or if I would have understood his missing our appointment, or if I would be meeting him still. The Saturday evening he had asked me to telephone I spent deep in thought, then suddenly I had the urge to see my night-school friend Songja[65]

It is during her walk around the fruit gardens at Oryu-dong with Songja the next morning that Nam-su reveals the depth of her resentment. Nam-su and Songja stumble across an enclosure of dogs that will be used to make soup. Songja explains to Nam-su that the dogs have been drugged so that their barking does not carry and disturb the neighborhood, and Nam-su stands and listens, transfixed by the dogs' hoarse voices and bulging eyes. This is her moment to reflect, which she does ruthlessly: "As I got closer I saw that the dogs were all barking in unison but no sound came from them. It is a dog's instinct to bark and yet here they had been muffled while they strained to bark in the most agonizing way. . . . I too felt my throat constrict and something seemed to grip my chest."[66]

Chang Nam-su recognizes herself in the trapped, gasping dogs. She describes the feeling of not only being trapped by class—by poverty, by lack of education, by her sensitivity to slight—but above all knowing better than anybody exactly how trapped and voiceless she is. Chang is compelled to reflect on the social contradictions that have thwarted her desires:

> The love stories that have come down to us from olden times were of a king and a country maiden, a princess and a woodcutter. But in this society it is university student and university student, the boss's daughter and the son of an elite government bureaucrat, the worker with other workers, and so it all seems to fit into a tidy cliché. It is not people meeting each other with the deep sense of [sharing something as] human beings, but name goes with name, prestige goes with prestige,

> and so it goes on. . . . When we who work in factories fall in love, we're called sluts. If a university student gets into a scrape it's viewed indulgently, but if a worker makes a mistake everyone is disgusted.[67]

Chŏng Hyŏn-baek interprets Chang Nam-su's relationship with Hyŏn-u as an opportunity to escape her working life and rise in class status through love and marriage: "As they become closer we can sense [Chang's] intellectual curiosity and the student's attraction to her intelligent attitude. With the deepening of the relationship she falls into a profound internal conflict. She ends up deciding that he cannot suit her based on the understanding that they are both people of their class. In so doing she throws away this opportunity given to her [to rise in class status] and reveals her resentment."[68] The implication is that Chang is refusing the transformation from factory girl to lady. She will not cross classes, and by this stage the reader can recognize the principle she is struggling for—the utopian vision of a different society where people can meet and fall in love, not as "factory girl" and "university student," but free from the constraints of a class society. In fact, for Chang, love exposes more directly than any possible argument why love cannot solve all—why politics, society and money must intrude.

Stories of slum romances that ended badly for factory girls abounded in the South Korean union movement. Cho Wha Soon recounts the tale of a suicide where the worker took her own life near the house of the student who had betrayed her: "Their loneliness leads them to men. Amongst those workers who dream of a love match with a university student this kind of ending is not uncommon."[69] Kim Seung-kyung relates that dating a university student was seen as "prestigious" in the Masan factories where she worked.[70] She reports that when the Yonsei University student newspaper carried a story claiming that factory girls from the Masan Free Export Zone were sexually exploited by male students from nearby Kyongnam University, the Kyongnam students reacted by burning an effigy of the reporter. Sensitivities on this subject were great on both sides, it seems. Novelists also took up this theme of class romance. In Pak Wanso's short story "Mother and Daughter at Tea Time," the factory girl's husband is a former *hakch'ul* from a prominent family and with a degree from "one of the best universities in the country," and she lives in fear that as he leaves his radical past behind he will grow estranged from her as well.[71] As Namhee Lee argues in a perceptive reading of this story, "She fears that it is not only their differing social and educational backgrounds that sets them apart, but also the fact that she is equal to him only when he is totally committed to the ideals and goals of the movement. . . . Feeling that only within the realm of movement ideals could she be truly respected by him, she tries to believe that he still retains movement ideals even when everything around him indicates otherwise."[72] The reciprocal desire of factory girls for the university students who had sought them out appears in literature as a problem that had no solution. That story after story ended badly for the factory girl helps explain Chang Nam-su's repudiation of a class romance also. If we read the coda to these relationships in factory girl autobiographies, we learn that optimism about the labor movement for a growing working class goes hand in hand with pessimism about class romance for factory girls.

FACTORY GIRL VIRTUE

Chang Nam-su's love tales tell us not only about the social world of factory girls but also about their sense of morality. The labor of factory girls that degraded them in the eyes of a patriarchal and class-conscious society also brought them into social proximity with women who labored in brothels in the booming informal economy of the 1970s and 1980s. The reticence of factory girls on the subject of prostitution is striking. Chang Nam-su's *The Lost Workplace* is a book determinedly about factory girl virtue.

It is clear that factory girls who struggled to convince the public that they were respectable could not afford to compare themselves to prostitutes. As Elaine Kim has noted, the "traditional [upper-class] Korean view that women's labor outside the home degrades the family" meant that when women from the poorer classes entered employment society in the new social space of the factory they were conspicuous as an affront to femininity.[73] Yet this focus on a lack of femininity is in many ways a mystification that distracts us from the extreme sexualization of lower-class women that went hand in hand with their economic vulnerability. How else could it be that these "unfeminine" factory girls are the same prey of Seoul's evening streets, crowded buses, palaces, and factory districts? Here the gap between femininity and sexuality is bridged by sexual harassment that becomes a form of sexual discipline to make "unfeminine" women more sexually docile.

Because womanhood was defined by attributes that were out of reach for the poorer classes of women, working women came to be seen as tainted by an unfeminine need for money. The received image of working-class women as tainted by economic need affected factory women acutely. Femininity, beauty, and grace were "class properties" in South Korea, unattainable for factory women.[74] Chang Nam-su explains:

> People say that women's voices should not go over the wall; women should be modest, talk in a cultured manner, and behave gently. . . . Then what are we? If we measure ourselves in those terms, we are nothing. We have to be loud in order to communicate on the shop floor; we have to wear uniforms and rush between the machines; naturally our movements are coarse. If the only price we get paid for our endless working for our country's industrial development and economic growth is the contemptuous name of "*kongsuni*," and the deprivation of our "femininity," then, what are we? For whom are we working, and for what are we living?[75]

But while their femininity was in question, working-class women's sexual availability became overdetermined in this particular type of "law and order" society that barely acknowledged sexual assault as a crime. This is the surplus unwritten context of factory girl autobiographies, a context that the union movement itself could not acknowledge. As Kim Won has shown, factory girls were constructed as either "sexless" or "productive labor" by a labor movement that had no capacity to critique the breach of both Confucian norms and middle-class morality that laboring women appeared to represent.[76] In this way these authors had much more at stake and addressed a far deeper set of problems for working women than the labor movement was able to encompass. The authors tell us how nerve-racking, thrilling, and fraught it was to look for love in a social world that read your class background at a glance and made its own assumptions.

BUT IS IT LITERATURE?

It is important to note that prior to the publication of factory girl autobiographies, works in this genre authored by working-class women were not part of any recognizable literary tradition in Korean society. Indeed when they first appeared, these books opened up the whole question of the production of cultural authority. The authors make explicit the connections between the conditions of literary production, the emancipatory role literature can play both personally and socially, the joys of writing, and the haggard schedule a working woman must obey if she wants to publish anything. They remind us how ill-equipped they are to join the canon, membership in which is conditional on one's prose being swept clean of the writer's sweat and other evidence of exertion.

That being said, the works by Sŏk Chŏng-nam and Chang Nam-su do not stand outside the literary market that published and received them. Indeed the most popular work at the time of its publication, Song Hyosoon's *The Road to Seoul,* relays a sensibility of Christian meekness and feminine suffering, a kind of "working-class weepy," that makes it the most dated of the three works today. More than the autobiographies by Chang and Sok, *The Road to Seoul* fit into already established conventions of the portrayal of working-class women as sentimental figures, in literature as well as in the labor movement. By contrast, Sŏk and Chang present themselves as literary and social subjects in ways that implicate the middle class in their experience of degradation. They illustrate the slippage between the poverty of their working life and the richness of their interior worlds, and Chang in particular clearly names the sexual politics of poverty.

Perhaps because of this these two authors remained ambiguous presences in South Korea's literary circles in the 1980s. We need only turn to Chang Nam-su's introduction to her autobiography to observe how difficult it is for her to place herself: "I am no famous politician, no celebrity, no artist. . . . Like most workers I am only the unwanted daughter of a poor farming family who became a factory girl."[77] Here she is longing for a secure place in the cultural and social hierarchy, while at the same time deriding that hierarchy. Perhaps one of the most striking innovations shared by these autobiographies was to tie political and literary representation together in works that are framed as a challenge both to literature and to society, foreshadowing a new kind of world where working-class women might be central protagonists.

It is no coincidence that the 1970s and 1980s produced our richest sources on the lives of female textile workers in South Korea. These were the years when women workers fought for and gained the attention of their society through a militant union movement and created a space for themselves in literature. For Chang Nam-su and other factory women, fighting low wages was also fighting the repression of desire. Chang's autobiography and works like it show how difficult it was to be sexually curious on small wages, how gender discipline encompassed by the word "femininity" worked to control factory girls, and how class and sexuality were inextricably intertwined. Despite these restrictions, Chang Nam-su writes of how one might reject low wages and degrading conditions and all the determinism of poverty and believe in the possibility of other kinds of relationships, glimpsed one night on a slow train to Seoul.

5. Girl Love and Suicide

"You've taken teenage workers as your theme," Yun-ho said, "and talked for thirty minutes or so. You talked as though you were knowledgeable, but you aren't. There isn't one person in our country who can talk about teenage workers without feeling guilty. And that includes me."

CHO SE-HUI, *The Dwarf*

Sometimes on the evening bus home or walking down an alley I would remember her. Can she really be dead?

SHIN KYŎNG-SUK, *Oettan Bang* (The Solitary Room)

This book has examined the process by which factory girls became cultural figures of immense political significance in modern Korean literature and its industrializing society. Though I have been primarily concerned with the politics of representation, this book's central argument pivots on the question:, what was the process by which a factory girl might become an author herself? I pose these problems of representation and authorship together because the gap that lies between them, the gap between being represented and fashioning the portrait, reveals a great deal about the simultaneous suppression of writing by working-class women and their power as literary heroines. In this chapter I examine a book that delineates the means by which a factory girl might become an author herself. That book is Shin Kyŏng-suk's *Oettan Bang* (The Solitary Room), which was first published in 1995 and is structured by a series of flashbacks that takes readers back to the late 1970s and early 1980s. *The Solitary Room* is about becoming a writer, narrated (and authored) by someone who spent her formative adolescent years in the factories of Kuro Kongdan.

Since its publication in 1995, Shin Kyŏng-suk's *The Solitary Room* has been appraised as the pinnacle of factory girl literature. Said to represent the culmination of seventy years of radical journalism, proletarian literature, and working-class autobiography, it is another in a long line of works that attempts to pin down the elusive subject of the factory girl. Both critically acclaimed and a best-seller, *The Solitary Room* takes readers slowly and gently back to the recent past of the brutal process of rapid industrialization. Locating its working-class women and girls in industrializing Seoul of the late 1970s and early 1980s, the book fixes on adolescent sexuality and suicide to tell the story of the emergent subject of the factory girl.

The book begins with the narrator one evening on Cheju Island looking out at the sea and remembering her fifteen-year-old self: "A chubby faced, ordinary-looking girl such as you could find anywhere in this country. It was 1978, the end of the Yushin period."[1] And so begins the story of a "teenage worker" or "factory girl." Sent along with her cousin to Seoul, the narrator—named Kyŏng-suk—continues her education the only way her family can afford it: as a self-supporting industrial high school student who is a factory worker by day and a schoolgirl by night. The main part of the book is set in the years 1978–81 and covers four years of the narrator's life, from the age of fifteen to nineteen. "I am not yet reconciled to those four years of my life," she declares repeatedly in the narrative. The book also includes the narrator's adult self, her successful novelist self, struggling to assimilate those adolescent years into her present life and to reconcile the Seoul of the late 1970s with the Seoul of the mid-1990s.

The book itself, with its "I" narrator, is part fiction, part memoir—a deliberate blurring of genres that the author confesses is her strategy to be able to write about these bleak years. The feminist literary historian Yi Sang-gyŏng has pointed out the slippage between the "I" of the narrator and the "I" of the author in *The Solitary Room*, given the revelatory, almost confessional tone of the text and the facts that are widely known about Shin Kyŏng-suk's life.[2] In her author's biography in previous books, Shin had recorded along with the barest details of her birth and early life the note that she had attended an industrial high school in Seoul. This one statement alerts readers that she had worked in factories while finishing high school, disclosing a whole world of adolescent hardship. Yet *The Solitary Room* is not an autobiography. In the final lines, the author writes, "This story is done and yet it seems to have been neither fact nor fiction but caught somewhere between the two. I wonder if it can be called literature. Or should we think of it as writing. And what is writing? I ask myself again." Right up until the end of the book the author grapples with form and genre and lays before the reader how reluctantly yet compulsively the book was written—how much peace of mind this book cost her, how it turned her into a negative, angry introvert.[3] So why was this story so hard to write? What is the function of the author's literary strategies? Who or what does she use to drive the narrative back into her past?

The author blends multiple genres in a narrative strategy that seems an open criticism of the conventions of factory girl literature. Thus autobiography and narrative fiction are brought to bear in telling her story; domestic and industrial fiction conventions influence the characterization of family and coworkers; and gothic disturbance is interlaced with distant memoir. If, as I argue, Shin Kyŏng-suk was in the process of formulating a subject that had not yet existed in literature, the impulse that she was writing against was not an invisible, as yet unrealized self that she had to conjure out of political obscurity, but rather the mass identity of the factory girl who represented in a series of clichés an entire generation of exploited laborers. We have already examined the establishment of conventions for writing about lower-class women, the tropes used to display them to political advantage, and the difficulties this tradition posed for the first women to take up their pens and write of their lives in the 1970s and 1980s. Shin Kyŏng-suk both deploys and resists these tropes. Her book hinges on what might seem to be the genre's ultimate cliché: the death of a factory girl. But whatever pleasure the reader might expect to take in the "aesthetically healing powers" of a narrative set in the poverty and loneliness of Seoul of recent memory is disturbed by the narrator's obsessive grief and anger and the emotions that drive the book: the pain of loss, guilt, and complicity.[4]

To navigate the delicate territory of thinking about how a subject is created in and through literary fiction, I borrow the insights of Nancy Armstrong. In her book *How Novels Think,* Armstrong reevaluates the history of the English novel as a history of the creation of modern individualism: "the history of the novel and the history of the modern subject are, quite literally, one and the same," she asserts.[5] Armstrong examines a body of fictional works as social agents engaged in mapping the limits of individualism. As writers sought to articulate a subject that had not yet existed in literature, they strained to reconcile individual desire and the constraints of the social order. When individuals marked by their excessive (asocial) desire die in canonical fiction (e.g., Maggie Tulliver in *Mill on the Floss,* Catherine Earnshaw in *Wuthering Heights*), it is to preserve the social order that such desire threatens to invalidate. For Armstrong the act of writing fiction creates the subjects we seek to be and enlarges the compass of the society we wish to inhabit. Writing, and the sympathy it requires of readers, constitutes a crucial element in subject formation. Thus the narrator of *The Solitary Room*'s experience of toiling in the factories only marks her as one of thousands; it is when she crafts a literary novel out of this experience that she becomes exceptional, a fully individualized creation.

Concerned with the role of literacy and of gender, and of the repressed, alternative subject of modern novels that competes with individualism, Armstrong covers territory that holds great interest for the reader of factory girl literature. Her final section examines the limits of the human community being outlined in philosophy as well as literature in the nineteenth century and takes readers into the gothic oeuvre of Bram Stoker's *Dracula,* Rider Haggard's *She,* and Mary Shelley's *Frankenstein.* Here Armstrong uses a history of feminist cultural theory to suggest that the modern individual is defined in terms that require the constant negation of alternative "idiosyncratic, less than fully human, dangerous" subjects.[6] *The Solitary Room,* a book about the suicide of one factory girl and the success of another, is deeply enmeshed in this project of creating interiority and selfhood out of a destructive and threatening past. The capacity of literature to confer individuality is not something easily accepted or endorsed by the narrator of *The Solitary Room,* however. The most important relationship in the book—the friendship between the two factory girls, the narrator and Heejae—slips in and out of the pages in tantalizing references, but we are on page 145 before they first meet.

"I think the first time I saw Heejae was the spring of 1978."[7] The narrator recalls one Sunday after returning with her cousin from the public baths, catching sight of a girl in their cement courtyard washing a school uniform. The crimson uniform is identical to the narrator's own: they must attend the same school. Their first words are spoken on the boarding house's roof as they hang up their washing. The narrator describes only Heejae's face: "a face innocent and unrevealing as sunlight," before her attention snags on Heejae's industrial injury:

> "Yes it's from a needle piercing me. When I put my hand in water it soothes it. Which room do you live in?"
>
> "I'm on the third floor."
>
> "I've seen you on the bus. I've seen you once at school too . . . but I didn't know you lived here."
>
> "I've never seen you before," I said.
>
> At that Heejae laughed softly again. She seemed to have assessed instantly that I was younger than her and she talked to me as though I was her younger sister. I found myself replying "Yes, yes" politely.
>
> "I like this rooming house. . . . You could die here and no-one would know. Isn't that so?" she asked me.[8]

So their friendship begins, shadowed by hints and portents. Here we get a sense of the immense appetite for female friendship, love, admiration, and worship that courses through the homosocial adolescent worlds of factory and high school. "She was gone, walking through the heavy sheets while the afternoon wind flapped and swirled like a curtain heavy with secrets."[9] In this scene of their first encounter, the boredom and thankless tedium of work and school are swept away in an instant of recognition.

To Kyŏng-suk, Heejae's intensity and sensuality are intoxicating. Kyŏng-suk reads Heejae as elusive, "as elusive as sunlight." But perhaps that is her warning to readers: do not imagine that you can comprehend the dead factory girl. Sympathy is not a currency that grants you entry into this world. You need first to understand that I bring you here *at great cost to myself*. That it is not my desire to make Heejae known to you, but to be clearer to myself. To write myself into being as the many things I am, simultaneously: innocent and complicit, grieving and successful, lonely and read by everyone.

As the story unfolds the narrator discloses more and more about the pain of remembering and the compulsion to write. She needs privacy. She needs in fact to escape her routine and flee alone to Cheju Island in order to return to this story and grieve and console her teenage self. The author relates the enormous cost, the ambivalence, the repulsion for going public with a private grief that she knows will be taken up as emblematic (of women's liberation), as part of the struggle (for workers' rights), as one chink in building "the greater good." Even as she writes the narrator is disconcerted by the process of publishing, of giving interviews, of scraping back this private death to show the world. The voice of Heejae haunts her, and perhaps she fears that by telling the world she will lose this "voice that lives on after death, so sweet and full of life and power."[10] Writing here begins to lose its luster, and literature feels like a mediocre compromise. How can literature compare with the voice of the beloved dead?

After pages and pages in which she searches for Heejae, finally the author reaches for the conventions of the gothic to bring Heejae into her present life. The present-day narrator is sleeping at home when she is awoken at 5:15 A.M. by the loud peal of her doorbell. She goes to the door but hesitates to open it and starts to wonder if she has imagined the sound of the doorbell. She returns to her bedroom:

> As my heart leaped around in my chest I felt the presence of a person behind me. I started with fright and turned around. My shawl that had been hanging on the back of a chair had fallen softly to the floor. I collected the shawl and gave a little relieved sigh
>
> . . . Then I felt that someone had come into my room.
>
> . . . I called out "Who is it?" But this person, whom I felt behind me, watching closely the nape of my neck, could not reply: "It is me."
>
> . . . I gave up and turned off the light and returned to bed. The shadow followed and curled up next to me.

"Heejae? . . . Is that you? . . . You gave me a scare. . . . How did you know where to find me? . . . Life has turned out well for me, hasn't it. . . . I'm so sorry.

". . . What did you say? . . . Heejae, what did you say? . . . I didn't catch it, can you speak a little bit louder? . . . What? . . . I can't hear you. What did you say?

". . . I write so that I might begin to touch you, Heejae."[11]

In a scene reminiscent of Heathcliff and Catherine in *Wuthering Heights*, Heejae resists death and literature also. Here Shin Kyŏng-suk invites us to listen to "the past, the ostensibly dead" as an alternative form of evaluating the brutal 1970s.[12] Throughout *The Solitary Room*, the author mediates between one of the dominant clichés of factory girl literature—the death of a factory girl—and conveying the real impact of her friend's death. In factory girl literature, it was the very suffering of lower-class women that made them politically important. Yet in the suicide of Heejae the narrator as mediator becomes the figure we fix on, the hidden protagonist set to reveal herself at last. Instead of projecting our own (bogus, bourgeois) emotions onto the act of self-annihilation, the reader must contend with the real impact this death had on Kyŏng-suk. It does not ask us to mourn (at no cost to ourselves) a distant, inaccessible life; rather its show us the effect on the living. The reader's vicarious grief is not solicited. It is the narrator's grief and anger and guilt that reveals itself as having been the narrative engine all along.

Heejae slips through the conventions of factory girl literature to reveal a new realism of female working-class life: Was she a worker or a student or a seamstress or a bar girl? Was she a teenager or considerably older? Did she kill only herself or her unborn child as well? And, finally, the narrator becomes the conduit between the reader and the inaccessible past, the remote class, the dead factory girl. Thus in *The Solitary Room* sentimental literature becomes realism.

WRITING, DESIRE, IDENTITY

The Solitary Room, a book about the struggle to become a writer and survive adolescence in the factory districts, raises the question of the disconnect between authority and representation. It is clearly not the case that factory girls were absent in literature prior to the 1990s. We might rather say that they were present in a way that aided in the suppression of their writing. They were represented as literary archetypes, as victims of sexual violence, exposed by poverty to the brutality of colonial capitalism in the

1920s and 1930s and by rapid industrialization in the 1960s through 1980s in ways that stifled them as creative figures. None of this is glossed over in *The Solitary Room*. When we open this book we encounter the full force of gender and class discipline operative in factories in the late 1970s and the ways that ambition and imagination are swallowed up by everyday labor. But Shin Kyŏng-suk alone is able to show that one form of silencing is connected to another—that the same trope of sexual availability and suffering that inhibits her narrator's confidence also injects dramatic power into her story. In *The Solitary Room*, as in all bildungsromans, the fate of the characters hinges upon how they manage their social and asocial desires.

The narrator forms her identity and tests it in the world by articulating what it is that she desires. She and her friend Heejae play a game in which they reveal their deepest wishes, the desires that will define them. Fantasy and brute economic reality are entwined in the adolescents' longings where ideal jobs as bank clerks and telephone operators compete with fantasies of love and motherhood. Heejae's innocuous question "And shall I have a pretty child?" is a part with the foreboding that shadows these exchanges. The mixture of fantasy and constraint that characterize these wishes is played out in an intense relationship of *chamae-ae*, or "girl love," that Kyŏng-suk and Heejae share. While Kyŏng-suk is tall, shy, and earnest, Heejae is intense, sensual, and elusive. Kyŏng-suk is sometimes shocked or at a loss for words in the face of Heejae's stories, but Heejae is the friend who knows how to comfort Kyŏng-suk and treats her with delicacy and care in the great heartbreaks of her adolescent years. Heejae is regarded with quiet suspicion by Kyŏng-suk's cousin and eldest brother; she lives alone and her age is indeterminate. Heejae is not easily slotted into the conventional markers of her working-class community, and she is apt to throw herself with gusto into love affairs. Yet her sensibility exactly suits that of the chaste narrator, Kyŏng-suk, and in their boardinghouse, and at the industrial high school they attend together, Kyŏng-suk and Heejae become intimate friends.

Despite being a new coinage, girl love was not outside convention or an unspeakable form of young women relating in 1970s South Korea. As Sharon Marcus has shown in her discussion of female marriage in nineteenth-century England, homosociality, rather than being a transgressive act, instead can be said to have ordered social and institutional life in bourgeois European culture.[13] In South Korea, girl love in the homosocial world of the factory and the high school is at the heart of normative institutions and discourses about family, love, and productivity. In *The Solitary Room* girl love reinforces the primacy of heterosexual partnerships even as it supplies the most intense relationship in the book. It is with girl love that we approach key questions of factory girl literature of the 1970s: Is the factory girl a sexual subject? What kind of sexual subject? And how does one desire in this environment?

Kyŏng-suk and Heejae's intense, confiding relationship is an expression of their desire and its limitations. The author gives us characters who connect most deeply in a homosocial world, yet it is an environment with no concept of homosexuality. This is significant because the author has already eschewed so many conventional markers of working-class subject formation. Again and again the author/narrator shies away from the factory demonstrations and union building that has traditionally established the contours of working-class subjectivity in this era. In her intimate turn toward interiority, desire, and girl love, the author presents an enduring homosocial relationship without ambiguity: it is completely heterosexual. On the one hand the relationship between Heejae and Kyŏng-suk is the most powerful and sensual relationship in the book, and yet same-sex desire or satisfaction is inconceivable in this story. While Heejae and Kyŏng-suk sleep together in Heejae's cramped and chronically untidy room night after night, telling stories of their disastrous relationships with boys, these are characters who do not have the capacity to desire or find fulfillment in each other. The question for us as readers is, why is this so?

AMBITION, EDUCATION, FAMILY

There is a famous, much-quoted scene in *The Solitary Room* of the author standing at a conveyer belt with a novel propped open beneath the machine, reading. The industrial high school that the author/narrator attended is understood by the narrator and her family as her sole connection to a future outside her class. Education, family, and sacrifice are deeply entwined in the narrative as the author explains how it was that she came to Seoul and set up house with her older brothers and female cousin. When she finished elementary school and was about to enter middle school, her second brother passed his high school entrance exam early and the family had to suddenly find the money to enroll them both in school. But in a classic trope of maternal sacrifice, her mother sold her ring to pay for another year of schooling. It is a scene that seems to signify the moment of her mother's final, complete abnegation of her own desires. The reprieve did not last, however. As the author was about to enter high school, her third brother passed his university entrance exam and her younger sister entered middle school. There was no money to send Kyŏng-suk to high school, so she had to go to Seoul with her eldest brother and try to enter an industrial high school there.

In her book Shin Kyŏng-suk spends some effort unraveling what the males in her family made of the decision taken to send her to a factory so young. She writes extensively about her two older brothers and their relationship with her and depicts in detail their suppressed anger that their younger sister must get her schooling via a factory. Their own privations fray at their nerves, but both brothers go out of their way to provide a modicum of comfort to Kyŏng-suk and their cousin. Although they share the one room at a lodging house, the brothers wait outside on park benches when they think the girls might be changing clothes. They purchase them food treats and inquire into their ambitions. The oldest brother arranges some fake identity papers for Kyŏng-suk so that she can enter a "decent" factory.[14]

Kyŏng-suk's father appears only briefly, but he too enriches our understanding of the patriarchal family. When Kyŏng-suk at fifteen first leaves home for Seoul her father closes up the shop and does not leave his room for three days. Repeatedly in this story the male family members take on guilt and responsibility for the life course of Kyŏng-suk. When Kyŏng-suk and her cousin start work at the factory, Kyŏng-suk's eldest brother takes them out for dinner. Kyŏng-suk observes her brother through the fumes of the grilling ribs, seeing that he is exhausted and unable to eat. His worry for them, and anger at what lies ahead for them, is expressed in food, buying things for them, giving them treats. The eldest brother, who looks after them when they come to Seoul, is a military conscript who is allowed to work as an office assistant rather than going to the training camps because financial circumstances have constrained him to provide for his family. This was one of the three categories that gained exemption from the compulsory three years of barracks training, the others being physical weakness and the threat to family lineage (being the only son of an only son). Of course connections were also useful in securing exemptions or lighter duties. When the cousins first come to Seoul, the eldest brother does not have lodgings and sleeps in the local government office where he is the night guard. He is studying law at evening school. Thus eldest brother uses all twenty-four hours of the day, sleeping and waking, to get ahead. The narrator says of him that even without regular meals his skin is pure white and his hands are clean and his shirts dazzling white. "He [cultivates] the look of someone who had never known the world's privations."[15]

Kyŏng-suk shares with her brothers the shock of arriving in Seoul and discovering that in this world they are considered "lower class." The advice the brothers give is leavened by a shrewd appreciation of the obstacles that face them all. Their counsel is both brutal and sympathetic, as in the following exchange between the cousin, who does not want to register to go to school, and the eldest brother, placing education within the context of aspirational marriage:

> "How can I be starting high school at my age?"
> "How old are you?"
> "I'm nineteen."
> "What's so old about that?"
> "It's old. My friends have all graduated by now."
> "So how long do you plan to be at the factory?"
> Cousin compressed her lips.
> "You like people calling you *kongsuni*?"
> Cousin closed her lips tighter.
> "If you don't go to school then you'll never escape this life."
> Even at that cousin kept her lips closed.
> "Is that what you want?"
> Cousin's neck began to tremble.
> "Is it?"
> "Everyone lives like this," cousin retorted weakly.
> "Who says everyone lives like this? You're the ones who live like this. Everybody else goes to school and then to college and are able to do what they choose to do in life."
> By this point cousin was nearly crying, but brother pressed on.
> "So you want to live here like this all your life?"
> "It's not 'all my life'! I'm earning money to buy a camera and get married."
> Brother smiled and his voice was softer when he asked, "Why a camera?"
> All this time I had stayed quiet and observed the two, but here I piped up:
> "She wants to become a photographer."
> "She's living in dreamland then," said brother. "Marriage is the same. If you work in a factory then you're going to end up marrying someone else from a factory, that's how it works. In this country if you want to live like a human being you have to go to school." Although brother put this to her brutally, cousin still did not relent. Brother yelled at her:
> "If that was all you wanted then why did you even come to Seoul? Why didn't you stay home and work at a factory there? If you're not here to go to school then pack up your bindle and go home right now!"[16]

People's apprehensions about the rigidity of class distinctions were sometimes overstated, but when social sets were formed around hometown, schooling, and work ties it was easy to feel locked into a very small subset of striving, self-supporting worker-students. We have seen how education was documented as a flashpoint for factory girls, many of whom marked the moment of prematurely leaving school as the moment they confronted class division and had to prune their aspirations into a choice between monotonous varieties of work. Education was thus a prominent subject of fantasy in the documentary sources on factory girls. It was also the first entry point for the industrial missionaries, and the students who followed them, in developing curricula for workers' night schools. Education was thus both sealed off and slowly cracked open for politicized factory workers who were drawn to the night schools and were receptive to the idea of continuing their study even at considerable cost to their physical well-being. But there were tensions in reconciling the emancipatory possibilities of education with the reality of working people's lives. Chŏng Mi-suk has reported that the loss of education, and the sense of losing all control over one's fate in the job market, meant that the Christian and more radical night-school movement schools become attractive for working people as both a repository of education and a challenge to the political status quo.

This is complicated in Shin Kyŏng-suk's story, where the union and the school are pitted against each other. Only nonunionized workers are allowed the privilege of attending school, and the narrator and cousin must choose between the two under the gaze of their unionized coworkers. Their guilt at choosing the school at first distracts them from the hard consequences of combining labor and learning in a single day. The combination of work and school meant that rest and leisure were almost completely swallowed up for the promised benefits of learning. It is hard to overestimate the value attributed to education in this society. In Korea the economy and social life had for centuries been organized around educational attainment, itself dictated by birth. Education meant entry into the professions; in the absence of strong guilds or unions to arbitrate the pay of the skilled trades, conditions were harsh. People went to great lengths to avoid this being their children's fate.

In *The Solitary Room* the narrator's mother does everything within her power to ensure that her six children receive an education. The narrator's eldest brother endures years of night school alongside other ambitious indigent students to gain his law degree. This brother becomes adept at juggling his multiple lives as student, worker, military conscript, and eldest son and brother to manage the household in Seoul. The narrator discovers that her mother is illiterate in a passage that relates how frustrated Kyŏng-suk was when, as a middle school student, her mother did not read the stories she wrote and gave to her. In fact Kyŏng-suk's mother successfully hid her illiteracy from her perceptive daughter, and it is Kyŏng-suk's younger brother who brings it to her attention when he tells her on a visit home that he is teaching their mother to write. Thus it is the illiterate mother who instructs her children in the value of education, even as she hides behind hymn books she has memorized, all to avoid exposing the limits of her authority and the brittleness of the fantasy that makes up the core of her ambition for her children.

Fantasy and concealment is a key motif in education and other classed pursuits in *The Solitary Room*. In fact the author/narrator finds it difficult to write about her mother who cannot read. She asks herself if she was ashamed, if she deliberately avoided seeing her mother's illiteracy, even as she and her older brother were deeply embedded and complicit in her mother's fantasy of education for them.[17] The narrator connects her own willful ignorance of her mother's illiteracy with her compulsion to conceal from the publishing world her own education pedigree: "For years I had avoided talking about my high school, avoided people from the same high school, buried all memory of it until Ha Kye-suk found me. When my first book came out and I had to give them something for the writer's bio, I put down the name of my high school and then dreaded being asked about it. I became like someone who needed to hide their past, like an extravert switched into an introvert."[18]

Finally, we need to connect the fantasy of education to the dry, prosaic pedagogy found in Yŏngdŭngp'o Industrial Girls High School. The snobbery of the day girls, coupled with the distance between the experience of learning that Kyŏng-suk longs for and the tedium of accounting and bookkeeping that is her actual experience of education leads her to revolt. And in her protest she is not alone. Kyŏng-suk shares a desk with Miso, who keeps a copy of a book by Hegel (it is never specified which one) stashed under her desk and she is found furtively reading it throughout the school year. When Kyŏng-suk asks jealously what it is about, Miso responds that she has no idea. "Mind your own business" Miso responds, who cannot explain why she is compelled to run risks to read it in class while not understanding it. Kyŏng-suk returns to this scene later: "Much later I thought back on this conversation Miso and I had about reading Hegel. Now I know what it meant. By reading this book I prove that I am different to you all. I show that I despise you."[19]

As noted in the previous chapter, we can see that the complicated fantasies around education and office jobs did not necessarily express a simple desire for schooling or for entering the white-collar labor market. If we take them at face value, the repertoire of careers that were within the realm of possibility appear to be an impoverished range of choices. But if we look closer at the response to the narrator's revelation that she wants to be a writer—the excruciating exchanges, the surprise of her interlocutors, the concern on her behalf—we begin to see how fantasies needed to be carefully bridled lest they estrange their owners from their own working-class community:

> In a big voice cousin announced, "She says she's going to become a writer."
>
> "A writer? You?" My brother turned to me in wonder.
>
> My disloyal cousin opened her eyes wide in surprise, "What? Is it some kind of secret?"[20]

ISOLATION AND SUBJECTIVITY

The atmosphere of *The Solitary Room*—redolent with isolation and loneliness, its protagonist overwhelmed by the poverty of Seoul—appears to be signaled by the book's very title. Even as the narrator describes the boardinghouse she and her Seoul family lived in, with its thirty-seven rented rooms, she cannot account for why her overwhelming memory is of isolation: "It sat in a labyrinth of streets, in earshot of the station that is a conduit for the countrywide line to Suwon. . . . Its one window opened onto a scene of crowds exiting the railway station, a nearby market with its hole-in-the-wall shops was visible, and the overpass to reach them." In the center of a bustling working-class neighborhood, with the sunniest aspect of all the tiny lodging rooms in the building, the narrator is unable to reconcile the reality of her lodgings with the indelible impression it left on her, who "never once returned to see it again. Not only the boardinghouse, not only our room; I had avoided returning even to that district of Seoul, whose streets I can picture in my mind as vividly as though it were a photograph I had gazed upon for years."[21]

The isolation and loneliness that permeates *The Solitary Room* is noteworthy for another reason. Demographically the narrator was far from alone. In the late 1970s tens of thousands of country girls were making the same journey to the factory districts of Seoul. The author introduces plenty of characters who, like her, are great readers (such as her classmate Miso), or have artistic dreams (the narrator's cousin, whose dream is to become a nature photographer), or thirst for achievement (her eldest brother the law student), or who possess remarkable ambition and talent (her third brother the dissident). Kyŏng-suk's loneliness might be attributed to the anomie of late-industrializing Seoul and the menace of poverty. Or it might be that, in the midst of the long workday and the night classes and the sexual harassment and the miserly wages and the prospect of losing all one's beauty and youth in this cycle of labor, there was a writer struggling to emerge. One does not need to elevate writing as a craft to appreciate the threat to the narrator's subjectivity. In the years between when the events in this book took place (or events very like them) and when the book was published, a mass labor movement moved into the working-class districts of Seoul.

So it seems that the author was also writing about being alone in a mass movement. Shin Kyŏng-suk's story unfolds at a time when the only way for a member of the working poor to become a subject in industrializing South Korea seemed to be via the labor movement project for workers' rights. Yet the author resists this avenue of subject formation. Her book is structured by flashbacks that again and again resist tropes about factory girls, union activism, and a hard-won education.

For example, the narrator and her cousin, early on in their time in the factory, are stuck in the middle of a battle between the company and the union for the hearts and minds of workers. They finally decide to join the union, even as they accept that this decision might force them to forego the chance to get an education. This section in the book appears to retell one of many dramatic stories about the arduous battles over unions and working conditions at female-dominated factories in the 1970s. Yet in the midst of this struggle, the narrator appears in flash forward, abruptly breaks off the dramatic tension, and with the security of the present tense looks back to ask, "What of those people, what of those battles. What of the several thousand people in that company? Surely some will have departed the world already by now through some misfortune."[22] This observation compresses the drama of the previous section into a distant nostalgia; high political stakes are replaced by reminiscence of individuals she never knew. Why does the author do this? Is there a trope she is resisting, a way of writing about factory girls and their union battles that appears to define this period and this class to the exclusion of an interiority that encompasses all this and so much more?

When the narrator turns to describe the scene on the factory floor, her production line, her coworkers and the functions they perform in the factory, she uses the term *pungsokhwa*, or "genre painting." *Pungsokhwa* alerts us to the self-consciousness of these descriptions, how difficult it is to create a straight description of factory girls when artists and writers have made of women's labor a picturesque piece of art. On some level this is also a coming-of-age story that disturbs the sympathetic popular attitude toward "poor friendless girls" in Seoul that resolved the harsh edges of class difference into romances about gender mobility. Such heroines had their virtue rewarded by escaping the brutish world of factory districts and tiny slum lodgings in a narrative logic that condemned those left behind as the real working class, those who deserved or were already too hardened by adversity. But *The Solitary Room* in fact sits in the gap between being trapped by factory work and escaping it. When the narrator's old school friend, Ha Kye-suk, finds her again through reading her novels, she accuses the narrator of wanting to keep a secret of her early life spent in the vocational school and factory.

When Kyŏng-suk first gets the phone call from Ha Kye-suk, she struggles to place the name until in a flash she remembers, "Ha Kye-suk of the plump cheeks and red bottom lip that she chewed in anxiety every afternoon as she crept late into class."[23] Ha Kye-suk worked for a company notorious for its enforced overtime that made her late to school every afternoon. But were they students or were they workers? The author recalls this being the unspoken question they shared:

> We talked often until one day Ha Kye-suk said, "You haven't written about us."
>
> At another stage she said the same thing.
>
> "I have read your books. The only one I haven't been able to read is the first one. It is difficult for me to get out to one of the big bookstores. And my local bookstore couldn't order it in. So that's the only one I haven't read. You have written a lot about your childhood, and about your university days, and about your romantic past but there's nothing about us."
>
> [Silence]
>
> "It was because I wanted to see what you had written about our life that I had sought out your books to read."
>
> As I didn't break my silence, Ha Kye-suk dropped her voice to a whisper and said my name.
>
> "You're not ashamed of that part of your life are you?"
>
> Tense, I shifted the telephone receiver to my other hand. Ha Kye-suk, who a minute ago had been chatting away merrily, misinterpreted my silence and said with disappointment, "Well, you're living a different life to us now."[24]

As the narrator wrestles with this charge that she is able to write about so much that has been painful, but not this, the book unfolds before us.

The result of this dilemma is instructive. Shin Kyŏng-suk does end up writing about her four years working at a factory by day and attending night school, about this hidden life, and the book becomes a best-seller, one of the most critically acclaimed works of the decade in South Korea, translated and praised all over the world.[25]

Although set in the late 1970s and early 1980s, *The Solitary Room* actually speaks most deeply to the zeitgeist of the mid-1990s. By the mid-1990s, many people had a stake in the representation of working-class women, and the book could expect a keen and critical array of readers. Yet the title, the narrative, defined by isolation and loneliness, was something that the guilt-ridden postindustrial South Korea of the 1990s was extremely sensitive to. It is important to recall that this book received favorable reviews in nearly every publication where it was appraised in South Korea and became a best-seller in a deeply politically divided society. What was it about this book that made so many people across the political divide receptive to it? What was the nature of this consensus? People asked very important questions of this book and its reception. Prominent literary academic Paik Nak-chung posed the question, does the narrator's story—with its yearning for healing—point to a change that has already taken place in Korean society, or the possibility of such a healing (justice)?[26] The questions here are, What sort of tale about factory girls can be enjoyed by everyone? What is the story that is being savored?

CLASS HYBRIDITY

One of the striking features of *The Solitary Room* is that, rather than conveying a straight narrative of working-class communities, it gives voice to the extraordinary fluidity and hybridity of class in South Korea. The author relates how a person might be one class in the village and another in the city; one class in her teens and another altogether in her thirties. How a person might rail against the class he discovers he belongs to and indulge in class dreaming or assume a status that he does not own. Characters in this book borrow an imagined identity and try it on for size, an act that literary historian Yoon Sun Yang has dubbed "blasphemous" in another context, for its violation of hereditary status principles around which social relations were structured.[27] Characters hold on to an old self that circumstances have forced them to slough off: such as Heejae wearing her school uniform long after she dropped out of the industrial high school. Other characters wear their disguises to give them comfort, to give flesh to the fantasy, or to gain access to the inaccessible: such as the oldest brother in a wig to hide his military buzzcut, teaching at a *hakwon*.

In *The Solitary Room* Seoul emerges as a city where the appearance of class was extremely powerful, where money or a college degree could lift you free from the menace of poverty and drudgery. Heejae continues to wear her school uniform even when she has left school and is working and learning the rules of a completely different social world. She is spotted coming home from her new job as a seamstress at four in the morning, wearing her school uniform. The uniform signifies that she has a secure place in this heaving, anonymous world of striving and losing. But it also signals the creative potential of anonymity that the new metropolis holds.

Shin Kyŏng-suk's book reminds us of Seoul in the 1970s and 1980s, ruled by one military clique after another, a society run on suspicion and paranoia, where connections are everything. In such an environment, where so much is groundless and illegitimate, the ability to assume an appearance of substance becomes all the more important. This is Jacques Rancière's "culture in disorder where the prevailing system was in the process of disruption."[28] Class fluidity has its political sources but also its internal ones. We might think of the many acts of identity borrowing and slumming as a kind of displacement, a "sequence of disguises, substitutes and transfers by which we learn to get along with this self in a complex social world."[29] *The Solitary Room* snags on these costumes and disguises. The narrator circles them fascinated, searching for clues to the characters' dreams and lies, their chief weapons in negotiating a class-ridden society.

BEAUTY AND INDUSTRIAL INJURIES

Making literature out of the suffering and available bodies of lower-class women takes a toll on the narrator of *The Solitary Room*. This is clearly a psychic cost to writing these experiences down and opening them up to the gaze of strangers. When so much had been written and spoken about "the workers," "the proletariat," and "factory girls," Shin Kyŏng-suk asks the questions, How can you fight the conditions yet admire their result? What is a worker away from her machines, her production line? What is a writer separated from her books, her reading, her own pen? How does a writer *become* in a factory, in tiny lodgings shared with three other bodies? When so much else is believed to come out of factories—friendship, productive wealth, the stirrings of revolution,—why not writers too? Here we come to the difficulty those governesses the Brontës knew about and might be said to have foreseen for later factory girls: "Rather than fill the [post of governess] in any great house, I would have deliberately have taken a housemaid's place, bought a strong pair of gloves, swept bedrooms and staircases, and cleaned stoves and locks, in peace and independence. Rather than be a companion, I would have made shirts, and starved."[30]

Charlotte Brontë shows us the popular middle-class belief that to be a (female) worker, a shirtwaister perhaps, is to be free of concerns about gentility and female propriety. Brontë elevates plain, free, manual labor (starvation) over genteel servitude. The distinction is a dramatic one but loses its force when we actually enter the female factories where one's labor must be thorough, unstinting, physical but also subservient, quiet, neat. A concern for beauty does not disappear in the factory. Quite the contrary, one fears the coarsening effects of welding and soldering all the more when the results of the work appear in the faces and forms of colleagues:

> Cousin said, "It's lucky we didn't end up in soldering."
> "Why?"
> "Look at number 13's face."
> I turned to snatch a glance at number 13, who had come through the vocation training school with us and joined the factory at the same time. Above number 13's head smoke from the lead curled above the soldering flame. In only a few months her face had taken on a yellow hue.
> "It looks like it could be lead poisoning," said cousin.
> My fifteen-year-old self would study my face in the toilet mirror. Our landlady had said, "Watch out. Drinking water straight from the yard tap will give your face that pale color." Above my white face the yellow face of number 13 floated past. I too thought it a blessing we had not been sent to soldering.[31]

Exposure to industrial disease or injury took many different forms. The left-handed girl who sits next to the narrator of *The Solitary Room* in her night-school class, An Hyang-suk, works in a small confectionary factory:

> One day I grabbed her hand but quickly dropped it. It was rigid as a piece of wood. I felt I had released her hand too rudely so I took hold of it again. Left-handed An Hyang-suk appeared to know what I was thinking and she laughed aloud.
> "It's from wrapping all the candies. My hand has become stiff."
> "How many do you wrap?"
> "About twenty thousand a day."
> [Silence]
> Twenty thousand candies. I could not picture it. An Hyang-suk reached over and took hold of my hand.

"Your hand is so soft. You must have it good in your company."
Her hand stroking the back of my hand felt like the sole of a foot.
"At first the job interested me," Hyang-suk said. "I didn't know there was such a job as wrapping candies. But after several days of twisting the plastic wrapper around the candies my hands started bleeding."
Her fingers were crooked.
"Now my hand has hardened so that it doesn't hurt anymore. But two years ago I lost the use of my right-hand fingers. That's why I write with my left hand now.
"You can't tell anyone that my fingers are bung.
[Silence]
"OK?"
I nodded my head.[32]

In this beauty economy, girls must hide their industrial accidents and mutilations. Not because working-class men will not have them. Men cannot be generalized about with such certainty and must be presumed to have their own individual response to young women in their circle. The 1975 film *Youngja's Glory Days,* set in Seoul's factory areas and red-light districts, is instructive here. Youngja loses her left arm in a road accident when she is working as a bus conductor. When she takes up another job as a hostess/prostitute, she has her own set of clients who are charmed by her as she is. And finally at the end of the film Youngja is married with a child to a working-class man whose body is also injured. Rather than showing them as exemplary objects of pity or charity, the movie depicts this couple as a normal, scarred working-class family.

But the beauty economy is not necessarily primarily about heterosexual interaction. The rules are set and the competition waged in a homosocial community that takes its most assured pleasure in disciplining its own members. Even in the largely sex-segregated worlds of factory and school, Kyŏng-suk's cousin wears her rouge and lipstick. She compares herself ruthlessly with the girls around her: the eldest brother's girlfriend, workmates in the factory, the glittering Miss Myŏng who works in the manager's office. Winning the admiration and envy of other females is what drives her.

When the narrator turns to examine the other industrial high school students around her, she observes how the sobriquet "teenager" is itself a misnomer, a term casually slung over a diverse body of women struggling to find a place in the labor and education markets. As the narrator, the youngest pupil in the school, looks closely at the faces of the "schoolgirls" at Yŏngdŭngp'o Industrial High School, she sees grown women of twenty-two, twenty-five, twenty-six done up in uniform and bobbed hair, their faces "soaked in fatigue."[33] Anxiety over age, beauty, and exposure to injury are all intertwined in the factory experience, which encompasses the crucial years of growth and eligibility for adolescents becoming (single) women.

SEXUALITY

Shin Kyŏng-suk has said that she doesn't think of her works as straight "female coming of age stories." "I wrote about time periods. Time periods I couldn't just let go by, that I felt stuck in."[34] And in *The Solitary Room* the narrator's sexuality and her relationship with her sexed body does appear to be overshadowed by forces that define the times. The narrator's first menstrual period arrives when she is caught up in a cycle of overtime night work.[35] The pain is excruciating and, as she bends over in pain in the factory toilets, the narrator needs to be informed by her cousin that it is not a stomachache she is suffering from but menstruation. In fact, much of the narrator's understanding of her sexed body is informed by the factory world.

When the company's notorious sexual harasser, Yi Kyejang, sets his sights on Kyŏng-suk, Miss Lee, the union delegate, has already warned them of his reputation: "If he wants you he tracks you down like a dog on a scent. And if you try to avoid him or brush him off then the abuse really begins. He's a real arsehole. Be on your guard."[36] Although Yi Kyejang had initially targeted Kyŏng-suk's cousin, unaccountably he switches his attention to Kyŏng-suk herself and one day she finds a present addressed to her with a note. It is from Yi Kyejang asking her to meet him at a *tabang* after work that day. Kyŏng-suk is walking home from work with her cousin when the following exchange occurs:

> "What's wrong with you?" [asks cousin.]
> "Me? Nothing."
> "Yeah, right," retorted cousin and then suddenly she lost her temper.
> "What the hell is wrong with you?"
> "What did I do?"
> "Are you saying you're not acting strange? You're following so close behind me that I can't even walk. Is someone chasing you? And look, you're shaking! You've been like this all afternoon."
> [Silence]
> "What is it?"
> I pulled out the present that Yi Kyejang had given me. . . .

When she had seen the stationery set and read the note Yi Kyejang had written asking me to meet him at Unha Tabang, cousin found a bin near the markets and hurled it all in. "What a bastard. Well, he can wait there for you till his heart's content."

But then the cousin has a better idea and decides to meet the harasser. Dragging Kyŏng-suk with her, the cousin goes to Unha Tabang and declares they will insist that Yi Kyejang buy them tea and dinner and beer. But when they see Yi Kyejang waiting in the *tabang*'s interior, sitting in a dim fog of cigarette smoke, the cousin decides to confront him:

"Do you know how old she is?"
[Silence]
"She's fifteen."
[Silence]
"And she hasn't started menstruating yet."
My fifteen-year-old self lurched in shock.
"Kyejangnim, do you have a younger sister?
"It's because Kyŏng-suk goes to school like my little sister that I just wanted to buy her a dinner, what's the shame in that Miss Pak?" said Yi Kyejangnim
"Our eldest brother is the one who buys us dinner, not you."
With that cousin grabbed my hand and we walked out.[37]

On their walk home through the evening market, the cousin enlightens Kyŏng-suk on how the factory she works in is also Yi Kyejang's quarry. Kyŏng-suk, who had taken Yi Kyejang's protest that he was "just buying her a dinner because she reminded him of his younger sister" at face value, needs to be dragged into this new knowledge that transforms her understanding of the factory floor.[38] But Kyŏng-suk appears to deliberately prefer to live in the fantasy world where women disappear from the factory floor for a reason one need not inquire into. Kyŏng-suk's gullibility, or deliberate ignorance, is itself a kind of compliance with the code of silence around sexual harassment. Kyŏng-suk's sexual curiosity is extremely circumscribed by a longing for ignorance, and she expresses the dominant tropes of the era. Thus, when we encounter in the book fictionalized accounts of the Tongil and YH labor disputes, and other demonstrations and strikes that in the popular imagination characterized the factory districts of the late 1970s, we find the same linking of political activity and rape that was used to intimidate politically active young women. Instead of asking why political activism exposes women to rape, the author appears to validate this fear of the strike, the sit-in, the protest, as a moment of extreme sexual vulnerability.

THE DEATH OF HEEJAE

When Heejae stops attending school, her life begins to slide, and soon after that an old boyfriend reappears in her life. We next encounter her when Kyŏng-suk's eldest brother, returning home early one morning, sees her in her school uniform, her mouth twisted in a snarl. It is 4:00 A.M. At once the narrator's cousin and brother jump to the same conclusion—there is only one conclusion if you have slipped free of the institutions of school and factory: the *sulchip* (bar).

Even when Heejae explains that she is working at a dressmaker's shop, the cousin and brother (and reader) wonder if she is telling the truth. With all their doubts, the protests of naïve, willfully blind Kyŏng-suk seem to frame Heejae as guilty of sex and evasion. Working-class sexuality is so overdetermined in this watchful neighborhood that innocuous signs can be read as downfall. When Heejae perms her hair her identity as a schoolgirl is lost for good, and her social world reads in the curls her downfall, so starkly is sexuality and its signifiers encoded into hair styles. Heejae stands in for the problem of young women floating loose in Seoul, unable to be constrained by either school or factory. The need to make money drives them. A drive whose desperation others more snugly placed in the economy cannot grasp. To be desperate does not mean to lose one's ambition or dreams; it is rather that those dreams begin to allow for some degree of degradation along with the triumphs.

Heejae's death demonstrates the randomness of tragedy. Her mistake is sleeping with a boy, the same act that in many other working-class districts would identify her as typical in this context marks her for destruction. This is the moral universe of *The Solitary Room,* which cannot fully account for the reality of sexual harassment on the factory floor just as it cannot account for a sexually active, and thriving, working-class female figure. A sense of doom overshadows Heejae as soon as she is spotted late one night and named by Kyŏng-suk's eldest brother as a fallen woman, a *sulchip* girl.

Here we are reacquainted in a slightly different form with the lack of a language of sexual desire in factory girl literature. Given that the 1970s and 1980s, the decades of the worst excesses of the military dictatorships, were a dangerous time for lower-class women (coded as available), it appears that to operate in society as a desiring woman was difficult in the extreme. The language of desire is attenuated by a larger script of innocence and virtue. It struggles to emerge but when it does, in *The Solitary Room,* it exposes Heejae to slander on the one side and self-loathing on the other. The innocence (ignorance) of Kyŏng-suk allows the reader to suspend judgment, or rather to assess with impartiality Heejae's part in her own death.

In their final scene together, Heejae stops Kyŏng-suk as she is walking out of their alley and asks for her help. Heejae is going away, "to the countryside, to visit my family," but has forgotten to lock her room. Since Kyŏng-suk shares a key to Heejae's room, Heejae asks her to lock the door when she returns home from school that evening: "She looks at me and asks can I please lock her door when I return. Latch it from the outside. It's not difficult. I ask why doesn't she run back and do it herself now? She says oh it can wait. There's nothing worth stealing. Those are her last words to me."[39] When Kyŏng-suk returns that evening, she turns the key in the lock in Heejae's door on her way up to her own room she shares with her cousin. She waits for Heejae to return from her holiday. The days pass. Heejae does not return. The room remains locked. Kyŏng-suk can still feel in her fingers the pressure of turning the key. She takes to sitting outside Heejae's locked door. No sound comes from the room. Sometime later the door is broken down and Heejae is discovered dead in her room. The police cannot work out how she killed herself and avoided detection with the door locked from the outside. The narrator remains silent.

And with this scene it becomes clear that the narrator's long struggle to tell this story, the long silence in the book, the haunting by Heejae, is not from Heejae's so-called downfall, pregnancy, or death but from the pain of complicity, the "fingers that remember," the author's entanglement with death and the fact that she lived to build a world of opportunities when her beloved friend died at the height of their shared poverty. In trying to grasp the success of *The Solitary Room,* how it resonated with so many readers, we discover the consensus at last. It is not the politics implied by a literature of the working class; it is the suicide, the death by self-loathing. This is the shared grief, the collective complicity. And traces of the moral universe that condemned Heejae remain in the social world we ourselves inhabit. We understand enough about our times, we share enough of the same moral lexicon to know we should fear for a sexually active factory girl presented to us in literature.

By contending with death but refusing to die, the narrator appears to be enacting feminist philosopher Luce Irigaray's injunction that "staying alive seems to me to be a part of liberation."[40] The narrator does not hide her difficulty with staying alive in the society she has helped to create. Rather than endorsing the world of South Korea's democratizing mid-1990s society in which she publishes her history, the narrative relates the relentless silencing (by the narrator's brothers and friend) or shallow and manipulative "celebration" (by women's magazines) of the narrator's earlier life, which cannot help but stand in for the lives of an entire generation of teenage workers. Juxtaposed with this endlessly available mass identity of "factory girl," Heejae remains elusive even in death. Perhaps this was the only protection the author could offer in the face of an obliterating cliché.

The Solitary Room also speaks to the cultural moment when, to paraphrase Nancy Armstrong, class sympathy stopped being an action and instead started to be a feeling.[41] By the mid-1990s the factory districts were almost completely emptied of *hakch'ul,* and people in South Korea were busy coming to terms with the rules of engagement of the bourgeois democracy they had created. If the 1990s were for historians and writers the era of revelations about the past, when people had won for themselves the capacity to look back on history and the traumas they had lived through, Shin Kyŏng-suk looks back to find her reading, writing, and striving adolescent self.

Shin Kyŏng-suk makes visible for us the exclusions that constitute a literature about factory girls. It is a work of "literary solitude," a term I borrow from Jacques Rancière,[42] that accomplishes a new kind of story on the bones of historic battles. For example, it subscribes to the theology of the fallen women but not its class politics, and it shies away from the mass politics of class conflict that it has outlived. Ultimately this distance from the solidarity markers that characterized factory girl literature of the 1920s and 1930s, and 1970s and 1980s, is a source not of instability but of canonicity. Harnessing the luxury of distance, *The Solitary Room* spoke to its times as no other work of proletarian literature managed to do, yet its themes are the same old saws of guilt and sympathy. If writer and reader are both engaged in the construction of meaning, and literature tells us much about ourselves, what does the enormous success of *The Solitary Room* reveal about South Korean society and its (new) readers in translation?

The book shares the same fear of class conflict exhibited by the great nineteenth-century industrial novels, even as it accomplishes the most complete interior portrait of a factory girl in literature that this writer has ever encountered. Displaying a fear of both revolution and class immobility, Shin Kyŏng-suk's book navigates the rupture between sympathetic literature and its object. But even as she does this, Shin insists that we understand the source of the power of her story. She writes not only about her own youth but signals an entire cultural archive lost to labor, illiteracy, and self-doubt. She ponders the stories of friends and acquaintances from her past who track her down. One day a woman who graduated from the same high school calls the narrator: "'When I saw your photo in the paper, next to your book, I told my husband that I went to high school with you. He had never believed that I went to high school, because you know we never had reunions or met again. But after seeing you he finally believed me. And how hard we worked to graduate too!' She laughed and its merry sound cut me to the quick."[43]

By at once highlighting and questioning the value of writing, the burden of ambition, and the wake left by suicide, Shin Kyŏng-suk allows us to partake of a classed knowledge painfully acquired. Yet by underscoring the solitude of this literary project, whose success relies on readers' intertwined sense of guilt and curiosity, she never ceases to question the nature of writing and the exclusions that constitute its power.

Shin Kyŏng-suk's *The Solitary Room* was the last great work of industrial literature in South Korea. Over the 1990s, women workers disappeared again from popular culture, this time assimilated into tales and images of middle-class professional women and their struggle for a place in employment society. These professional women, whose stories were recounted in the smash hit television dramas of the mid- to late 1990s—*Son and Daughter, Cinderella, Chinshil*—did not often acknowledge that working-class women had been there before them, engaged in much the same sorts of struggles for recognition in the workplace. But while Korean factory girls may have disappeared from literature, they have not vanished. As human geographer Ayami Noritake has shown, some of the factory girls of the late 1970s and early 1980s became street entreupreneurs or ran garment workshops in the intervening years and have found a degree of autonomy in managing their own businesses in the old working-class district of Tongdaemun.[44] While the manufacturing sector has declined inside South Korea and business has relocated to China, Indonesia, and Vietnam, a new generation of women and men labor within the globalized South Korean economy. The form their stories may take, and the nature of the curiosity that may greet them, remains an open question.

Epilogue

In the classics of Korea's industrial literature, the love stories of factory girls are always shadowed by the threat of sexual violence. So interlocked are these narratives that desire itself became hazardous in factory girl literature. These stories interacted with the reality they sought to represent in ways both daring and furtive. Faced with the immense scale of the violence that went into implementing rushed industrialization, authors struggled to specify the nature and causes of this violence as it related to women. The strategy to encode and mystify sexual violence in factory girl literature was at once emancipatory and a measure of capitulation. As this book has argued, the parsing of the messy politics of sexual harassment, intimidation, and complicity in colonial-era factories as rape by one's boss saw gender politics completely absorbed by the discourse of class struggle. In the literature of the 1970s, the search for sexual pleasure in a menacing social world was coded as a journey that restated the importance of self-worth. In a society where violence against working-class women had no traction whatsoever as a political issue, factory girls were to find all the resources they needed within themselves.

That young laboring women's search for sexual pleasure should be hindered by their circumstances, as well as the language available to them, is underlined when we consider the volume of sexual violence unleashed by South Korea's particular version of militarized industrial development. While there was a midnight curfew for the entire population, the tacit curfew for young women was several hours earlier; as the evening wore on, trains and buses emptied of women and girls steering clear of the late-night commute. The architecture of rapid and shoddy development produced an endless array of dangerous sites. Seoul's suburbs were crisscrossed by unlit alleys. Everywhere women went, cautionary tales followed them. In restaurants and cafés female friends went to the toilets together to keep watch outside the cheaply constructed unisex booths where you exited straight into a urinal. And the sex industry, the liquor industry, all kinds of illicit trade was booming all around. Rape reportings increased 100 percent in the decade leading up to 1974 and doubled again within six years.[1] What we know of the 1970s and 1980s—the rapid development of the economy, the growing enrichment of the middle class, the shift from labor-intensive to technology-intensive capitalist growth, the expansion of education and ideas critical of the political and economic status quo—to all this can be added a general culture of sexual violence. Rather than targeting women only, this culture of violence functioned in multiple hierarchical relationships in the military, schools, employment society, and the family, which all preached to its objects endurance, respect, and silence.

Tracing the lives of Korean working-class women and their representation in literature, we can see that sexual violence is constantly referred to in ways both coded and uncoded, yet no clear analysis of its place and role in working-class communities exists. It appears that sexual violence was too ubiquitous to be noteworthy, too messy to be attributed to clear causes, yet nothing quite fits together until we see how regimes of sexual violence disciplined whole swathes of the population, especially under the military presidents. In an environment where harassment was without censure, and assault could make you a pariah among women, a regime of sexual violence did vital work in keeping a large segment of the population watchful, isolated, and besieged. The demonstrations at Tongil and YH were some of the very few occasions where this regime found it necessary to display itself openly. It should not surprise us that the austere military government of Park Chung Hee licensed its police to act in this way. As feminist academic and activist Insook Kwon has taught us, rape and sexual assault was simply another way that security agents expressed their loyalty to the state.[2]

That this essential ingredient of Korea's much studied industrialization should have slipped our notice is worth pondering. In the context of the ongoing reevaluation of the Park Chung Hee era in South Korea, do we include sexual violence in the logarithm that calculates the net losses and gains of Park's industrialization strategy? And in the same spirit, if we ask what the function was of all that violence, what answers might come back at us? Aside from intimidation, the outcome appears to be only endless amounts of waste and loss. But we must be sure to provide enough space to account for the state that allowed the violence and for the people that practiced it. Indeed, where does one end and the other begin? And here it is that literature intervened. By giving us character as well as context, blame and exoneration, crafting narratives from rumor as well as experience, and publishing open secrets, authors of factory girl literature revealed what was yet unnamed in Korean society.

And here at last we come face to face with literature's potential to aid or hinder self-knowledge, to spin further mystification or to name what is yet unnamed in our social organizations. Yet that the poignancy of factory girl stories should be based on the humiliations these women endured as newcomers to Seoul's relentless drive to industrialize also raises questions about their readers. What does it mean to read about the pain and degradation of another class? Is it an unmasking of an abusive world or a way of keeping alive the exhilarating *différence* of class? Is it an act of class sympathy that ultimately savors the distance between the protagonist and ourselves? The longevity of the factory girl character in modern Korean literature, in all her various eras and guises, speaks of readers' curiosity with this figure and the world she inhabited. In the 1920s and 1930s she was depicted firmly within the factory's economy of sexual violence and complicity, teaching readers about the treachery of the labor market and the seductions of capitalism. In the 1970s and 1980s she elaborated on female working-class virtue as a defensive strategy. In the same way that nudity was substituted for political demands at the Tongil Textile Company in 1976, in factory girl literature the open avowal of virtue substituted for, but never completely replaced, the search for sexual pleasure. In this way, in Shin Kyŏng-suk's *Oettan Bang* (The Solitary Room), the 1990s seem to have produced a book that is both more comprehensive and more conventional than earlier genres on this theme. Fascinated with all that conspires to ruin a factory girl, Shin Kyŏng-suk presents readers with a piece of empathetic fiction that partakes of many of the old fears about lonely working women.

Factory Girl Literature has argued that our understanding of Korea's rapid industrialization experience is incomplete if we fail to see how deeply sexual violence regulated the lives of the women who went out to work in the factories. More than simply establishing this fact, this book traces the coding of sexual violence in factory girl literature. The mystification or blurring of the politics of sexual violence, or its recoding as something else, does not mean that readers missed this plot cue. On the contrary, readers were fully equipped to understand the measure of what was unfolding before them. By relying on the shared secret of sexual violence that lay at the heart of rapid industrialization society, authors and readers were navigating a censorship that was also deeply productive. Enthralled by shared secrets, sensitive to nuance and coding, hungry for poignant tales that might confer meaning on meaningless suffering, thus did factory girl literature have its heyday.

Notes

INTRODUCTION

1. Clark, "Politics of Seduction in English Popular Culture, 1748–1848," 50.

2. Virginia Woolf's full quote is this: "Towards the end of the eighteenth century a change came about which, if I were rewriting history, I should describe more fully and think of greater importance than the Crusades or the Wars of the Roses. The middle class woman began to write." Woolf, *Room of One's Own*, 84.

3. Yi Ok-ji, *Han'guk Yŏsŏng Nodongja Undongsa*, 38.

4. Williams, *Culture and Society*, 104–5.

5. Here I am paraphrasing Jacques Rancière's critique of Pierre Bourdieu's *Distinction: A Social Critique of the Judgment of Taste*, found in "The Sociologist King," chap. 9 of *Philosopher and His Poor*.

6. Yi Sŏng-hŭi, "Hyŏndae-ŭi Yŏsŏng Undong," 397.

7. For an account of women's organizations in South Korea after the end of Japanese colonialism in 1945; their split into left- and right-wing camps, where they were overshadowed by male political organizations; and the triumph of conservative women's organizations under the patronage of the United States' military-occupation government, see Mun Kyŏng-ran, "Migun Chŏnggi Han'guk Yŏsŏng Undong-e Kwanhan Yŏngu."

8. See, for example, Kim Kyŏng-il, *Ilcheha Nodong Undongsa*, 3; and Chŏng Hyŏn-baek, "Yŏsŏng Nodongja-ŭi Ŭisik kwa Nodong Segye," 118.

9. The terms "proletarian literature" and "labor literature" refer to different time periods. In Korea, proletarian literature first appeared in magazines and newspapers from the mid-1920s, at a time when proletarian literature was a global phenomenon, popular in Japan, the Soviet Union, the United States, and Europe. Central to early formative debates on the nature of "modern" Korean literature in the 1920s and 1930s, proletarian literature from this period would find a new readership when radical presses in South Korea republished it in the late 1980s. Labor literature, by contrast, grew out of the South Korean labor movement of the 1970s and 1980s, when publishers,

authors and filmmakers found a new audience for a resurgent working-class culture.

10. Michael Denning uses this phrase in *Cultural Front,* 228, to describe the disappearance of promising proletarian writers in America as the 1930s progressed.

CHAPTER 1

1. The article is by an anonymous lady journalist *(puin kija),* in *Shin Kajong* 3, no. 2 (1935): 33, quoted in Lee Hyo-chae, "Ilcheha-ŭi Yŏsŏng Nodong Munje," 156. Women employed in the rubber factories were usually considerably older than factory girls in textile companies and silk-reeling factories, often around thirty years old, and it was not uncommon for them to work with their babies strapped to their backs. For more on this, see Yi Ok-ji, *Han'guk Yŏsŏng Nodongja Undongsa,* 38.

2. W.S. Park, *Colonial Industrialization and Labor in Korea,* 25; Lee Hyo-chae, "Ilcheha-ŭi Yŏsŏng Nodong Munje," 144.

3. Eckert, *Offspring of Empire,* 9–10.

4. For a discussion of the significance of 1919 as the beginning of systematic capitalist industrialization in colonial Korea, see ibid., chap. 2.

5. Ibid., 16–17.

6. H.K. Lee, *Land Utilization and Rural Economy in Korea,* 230. Wages paid for farm work were far below factory wages; even so, work was so scarce in the 1930s that, according to Lee, in the countryside there were "always crowds of laborers waiting for jobs" (229).

7. W.S. Park, in *Colonial Industrialization and Labor in Korea,* chap. 3, discusses how wartime changes in the cement industry in Korea affected both the labor force and the structure of industry. Carter Eckert discusses the expansion of key wartime industries in Korea in the late 1930s and early 1940s, in "Total War, Industrialization, and Social Change in Late Colonial Korea," 3–39.

8. Cumings, *Origins of the Korean War,* 13.

9. Kim Kyŏng-il, *Ilcheha Nodong Undongsa,* 37.

10. Kim Dae-hwan, "Kŭndaejŏk Imgŭm Nodong-ŭi Hyŏngsŏng Kwajŏng," 64. In addition to internal migration from the countryside to urban centers, there was also a substantial amount of emigration out of the country for work. In Japan, where labor had to be imported from the colonies in the 1940s to make up for the wartime shortfall, Koreans were mobilized to work in mines, factories, sea ports, and in the war industries. Bruce Cumings has suggested that more than one million Koreans were repatriated to South Korea from Japan between October 1945 and December 1947. Cumings, *Origins of the Korean War,* 60.

11. Quoted in Kim Kyŏng-il, *Ilcheha Nodong Undongsa,* 59 (my translation; all translations are mine unless otherwise noted).

12. Lee Hyo-chae, "Ilcheha-ŭi Yŏsŏng Nodong Munje."

13. Ibid., 144; Sin Yŏng-suk, "Ilche Singminjiha-ŭi Pyŏnhwadoen Yŏsŏng-ŭi Salm," 318.

14. Yi Ok-ji, *Han'guk Yŏsŏng Nodongja Undongsa*, 35. There were important continuities with women's work in the premodern period as Korean farm women, and increasingly in the nineteenth century impoverished *yangban* wives, took responsibility in the household economy for raising silkworms and spinning. See Yi Sun-gŭ, "Chosŏn Sidae Yŏsŏng-ŭi Il-kwa Saenghwal," 206–8.

15. So Hyŏng-sil's master's thesis, "Singminji Sidae Yŏsŏng Nodong Undong-e Kwanhan Yŏngu," compares industrial militancy by women workers in the rubber industry and the raw silk industry, documenting the structural conditions that impinge on the women's capacity to rebel. So Hyŏng-sil finds that, where women workers were able to combine and agitate collectively for higher wages in the rubber industry, workers in the silk factories were paralyzed by their isolation, harsh conditions, and the industry's culture of obedience. The high number of child workers in the silk factories may also have contributed to the degree of coercion that supervisors were able to apply to factory hands.

16. Yi Jŏng-ok, "Ilcheha Kongŏp Nodongeso-ŭi Minjŏk kwa Sŏng," 275–76, cited in Yi Ok-ji, *Han'guk Yŏsŏng Nodongja Undongsa*, 35.

17. Hagen Koo uses this phrase in describing Korea in his article "From Farm to Factory."

18. Eckert, *Offspring of Empire*, 193.

19. Sin Yŏng-suk, "Ilche Singminjiha-ŭi Pyŏnhwadoen Yŏsŏng-ŭi Salm," 318.

20. Mackie, *Creating Socialist Women in Japan*, 76. Vera Mackie makes the important point that the factory legislation concerning women workers, such as maternity and night work provisions, were framed in terms of the "protection" of females, not in terms of the rights of women workers.

21. For more on this, see Eckert, *Offspring of Empire*, 191–92.

22. W.S. Park, *Colonial Industrialization and Labor in Korea*, 114. Park reports that Japanese residents in Korea increased from 527,000 in 1930 to 713,000 in 1940. The presence of a considerable number of Japanese workers in the colony increased competition for jobs and pushed Korean workers into the lower, unskilled section of the labor market. Ibid., 21.

23. Ibid., 118.

24. Tsurumi, *Factory Girls*, 151.

25. Eckert, *Offspring of Empire*, 198.

26. J. Kim, *To Live to Work*, 88.

27. Eckert, *Offspring of Empire*, 197–99.

28. Lee Hyo-chae, "Ilcheha-ŭi Yŏsŏng Nodong Munje," 131–32.

29. Song Youn-ok, in "Japanese Colonial Rule and State-Managed Prostitution," demonstrates how the Japanese colonial state "modernized the sex industry in Korea under the state-run licensed prostitution system, which regulated business activity in the pleasure quarters." Song links the expansion of the prostitution industry with the enlargement of the war in the 1930s

and early 1940s, and she connects the prostitution-licensing system with the mobilization of "comfort women" by the Japanese military in the 1940s.

30. Quoted in Kang Yi-su, "1930 Nyŏndae Myŏnbang Taekiŏp Yŏsŏng Nodongja-ŭi Sangt'ae-e Taehan Yŏngu," 141.

31. See ibid., 143. In her dissertation, the social historian Kang Yi-su interviewed women who worked in the cotton-spinning industry in the 1930s about recruitment practices. She reports that "most factory employees remember the recruiters as 'unscrupulous people' *[chili anjoun saram]*. Aside from indulging in outright lies in the course of engaging [employees], it soon came to light that recruiters were involved in such socially abhorrent behavior as sexually assaulting factory women, or with the bait of finding them legitimate work in factories luring them to work in taverns or brothels" (143). Kang cites articles in the *Tonga Ilbo* in 1926 and 1927 that reported on such cases; see ibid., 143n37.

32. Kim Kyŏng-il, *Ilcheha Nodong Undongsa*, 567.

33. Ibid.

34. *Kaebyŏk* 4, no. 68 (1926), (translation by Chung Jin-ouk and Ruth Barraclough).

35. Lee Hyo-chae, "Ilcheha-ŭi Yŏsŏng Nodong Munje," 165.

36. Kim Kyŏng-il, 550.

37. *Tonga Ilbo*, November, 3, 1929. Yi Sŏng-ryong's letter of appeal dominated this issue's regular column "Chigŏp Puini Doegi Kajji" (Becoming Career Women), which ran in the *Tonga Ilbo* through 1929 and featured letters from a variety of women in the employment market, including *kisaeng*, teachers, and factory girls.

38. The term "sexual harassment" *(song hŭirong)* first appeared in public parlance in South Korea in the early 1990s, and sexual harassment only became an offense in 1999. I am grateful to Yi Eun-sang for this information.

39. "Foreign rice" here is presumably long-grain rice imported cheaply from South Asia. The detail given to describing dinners was a feature of writing by factory girls and displays the preoccupations of the hungry. This is also a characteristic of Korean labor literature from the 1970s and 1980s. The paltriness of the workers' remuneration can also be gauged by the food they were given—it was common practice for the cost of meals distributed to factory girls to be deducted from their wages, thus the poorest paid could only afford the cheapest meals.

40. Quoted in H. Lee, *Virginia Woolf*, 17.

41. Kim Kyŏng-il, *Ilcheha Nodong Undongsa*, 559–60.

42. Ibid., 560.

43. Information about this strike is taken from Lee Hyo-chae, "Ilcheha-ŭi Yŏsŏng Nodong Munje," 162–63; Kim Kyŏng-il, *Ilcheha Nodong Undongsa*, 531; Yi Ok-ji, *Han'guk Yŏsŏng Nodongja Undongsa*, 42–43; and coverage in *Tonga Ilbo*, July 3–16, 1923.

44. The title of one lecture was "From the Vanguard of the Working-Class: Chosŏn's Factory Girls." Lee Hyo-chae, "Ilcheha-ŭi Yŏsŏng Nodong Munje," 163. Lee also mentions that the women's supporters included union groups in Japan.

45. *Tonga Ilbo*, July 5, 1923.

46. This is not to suggest that writing went uncensored in this period. Rather, in the context of stringent censorship and intermittent arrests, writers navigated the literary terrain with skill and sensitivity, writing fiction, poetry, and essays in an atmosphere of heightened political awareness, or what J. M. Coetzee calls "over-reading." Coetzee, *Giving Offense*, 112.

47. Michael Robinson details the changes in Japan's governing elite that also led to this change in tactics in Korea. See Eckert et al., *Korea Old and New*, 276–85.

48. Scalapino and Lee, *Communism in Korea*, 66.

49. Eckert et al., *Korea Old and New*, 283.

50. Robinson, *Cultural Nationalism in Colonial Korea, 1920–1925*, 114.

51. See Kim Kyŏng-il's appendix to *Ilcheha Nodong Undongsa*,, 521–67. Indeed, these two newspapers were far more assiduous in covering labor disputes in the 1920s and 1930s than they would be in the 1970s.

52. From "The Diary of a Young Socialist," *Chosŏn Chikwang* (Light of Chosŏn), no. 68 (1927); also quoted in Kim Kyŏng-il, *Ilcheha Nodong Undongsa*, 427.

53. Sheila Smith has written about the Victorian novelists who in the 1840s and 1850s turned their attention to "the poor," whose alien lives they described in great detail to the higher classes to whom they had always been invisible. She calls this "the novelists' sensuous reaction to the poor"; writers like Charles Dickens and Elizabeth Gaskell wrote the physical detail of the lives and habits of the poor, whom they described almost as though they inhabited a foreign country. S. Smith, *Other Nation*, 23–44.

54. An example of such graphic depictions is the following description of a cotton-spinning factory in Pusan, taken from the July 2, 1936, issue of the *Chosŏn Chungang Ilbo*: "Female workers, usually between the ages of fifteen or sixteen and twenty, and most of them recruited from their villages in the provinces, work in this dark, dingy factory under the threatening surveillance of an overseer. The young women drink in warm air while the temperature sits near 100 degrees, their bodies throbbing, working until their bones crumble. They earn a maximum of 15 to 16 chŏn a day and work in this environment for six or seven years, enduring arduous training so that they can eventually be qualified to receive 40 to 50 chŏn a day. They live in what passes for a 'dormitory,' ten girls crammed into a room. The guard constantly moves them to make sure he can spy on them properly, and their freedom is extremely restricted. The workday is long, the food is indescribable, and these women's nutrition situation and health is extremely weak. Their pallor is like that of a patient at the end of a long illness, their bodies are emaciated and cases of fainting are frequent in the factory. [The factory] has rules and offending against these rules, even slightly, brings down immediate punishment by caning. That is the state of things." Other reports on working conditions in the factories include an article on the conditions of child workers, *Tonga Ilbo*, June 25, 1927, and extensive coverage of strikes, particularly those at Japanese factories. See

the July 1923 issues of the *Tonga Ilbo* for detailed coverage of the Kyŏngsŏng Rubber Strike discussed in the chapter.

55. Quoted in Kim Kyŏng-il, *Ilcheha Nodong Undongsa*, 61.

56. W.S. Park, *Colonial Industrialization and Labor in Korea*, 81.

57. Song Youn-ok, "Japanese Colonial Rule and State-Managed Prostitution," 190.

58. Kim Kyŏng-il, in *Ilcheha Nodong Undongsa*, 553–55, suggests the figure of 954.

59. Clark, *Women's Silence, Men's Violence*, 13.

60. Ibid.

61. This summary is taken from Catherine McKinnon, *Sexual Harassment of Working Women*.

62. So Hyŏng-sil, "Singminji Sidae Yŏsŏng Nodong Undong-e Kwanhan Yŏngu," 59.

63. Kim Kyŏng-il, *Ilcheha Nodong Undongsa*, 59–60, reports that in 1927, when the father of a fourteen-year-old factory girl at the Katakura Silk Spinning Factory in Taegu charged a factory supervisor with assaulting his daughter, the supervisor was granted a stay of prosecution.

64. For more information on such grievances, see Lee Hyo-chae's list of strike demands in the rice mills that relied on female labor. Lee Hyo-chae, "Ilcheha-ŭi Yŏsŏng Nodong Munje," 165.

65. Elaine Kim, in "Men's Talk," discusses this lack of social sanction for women's factory work.

66. Kang Yi-su, in "1930 Nyŏndae Myŏnbang Taekiŏp Yŏsŏng Nodongja-ŭi Sangt'ae-e Taehan Yŏngu," 222–23, describes one factory, the Chŏngyŏn textile factory, that was constructed like a prison— it had six observation towers to prevent escape, and those factory girls who did abscond were rounded up at the local railway station. See also Yi Ok-ji, *Han'guk Yŏsŏng Nodongja Undongsa*, 53.

67. Han'guk Yŏsŏng Yŏnguhoe, *Han'guk Yŏsŏngsa*, 241. The writer goes on: "This was also the case for the women's movement's largest body, Kunuhoe. Established by intelligentsia women it too failed to include the growing numbers of women workers, as well as peasant women, into its organization."

68. Lee Hyo-chae, "Ilcheha-ŭi Yŏsŏng Nodong Munje."

69. See Kim Yun-hwan, *Han'guk Nodong Undongsa;* and Kim Kyŏng-il, *Ilcheha Nodong Undongsa*.

70. Yi Ok-ji, *Han'guk Yŏsŏng Nodongja Undongsa*, 39.

71. Ibid.

72. Ibid., 53.

73. For laudatory accounts of the revolutionary and democratic possibilities of Chŏnpyŏng, see Ogle, *South Korea*, 8–12.

74. Yi Ok-ji, *Han'guk Yŏsŏng Nodongja Undongsa*, 62. The exception was socialist Chŏng Jong-myŏng, who was head of the women's bureau. Yi Yi-Hwa, *Han'guk Kŭnhyŏndaesa Sajŏn*, 238.

75. Yi Ok-ji, *Han'guk Yŏsŏng Nodongja Undongsa*, 62.

76. This was particularly the case for women laboring in Japanese mines, who

could be crushed by the hoists operating in mine shafts. For an account of Korean women miners in Japan, see W. D. Smith, "The 1932 Aso Coal Strike," 102–3.

77. Chakrabarty, *Rethinking Working-Class History*, xi.

78. The phrase belongs to Michael Denning, who used it to describe the dwindling of the proletarian-literature experiment in America in the 1930s. Denning, *Cultural Front*, 228.

79. See the editorial in the *Tonga Ilbo*, January 4, 1934. With this editorial, the *Tonga Ilbo* appears to declare itself once and for all on the side of the critics of the factory system, despite many prior years of even-handed coverage.

80. Denning, *Cultural Front*, 239.

81. It should be noted that illiteracy was not the condition of all females. In 1938 women made up 14 percent of teachers at Normal Schools. Lee Hyo-chae, "Ilcheha-ŭi Yŏsŏng Nodong Munje," 146.

82. The implication is that he was arrested.

83. *Shin Yŏsŏng* [New Woman], December 1931, reprinted in Yi Sang-gyŏng, *Kang Kyŏng-ae Chŏnjip*, 800–801. The word in the last line was censored.

CHAPTER 2

1. Ch'ae Manshik, "P'alyŏkan Mom," 45–46.

2. S. Smith, *Other Nation*, 2.

3. Yoon-shik Kim, "Phases of Development of Proletarian Literature in Korea." For a comprehensive account of the proletarian literature scene in Japan in these years, see Bowen-Struyk, "Rethinking Japanese Proletarian Literature," chap. 1.

4. *Kaebyŏk* was the leading progressive journal in colonial Korea during its 1920–26 run, after which it was shut down for its increasingly radical line. The Korean Proletarian Artists Federation, writer Song Yŏng wrote proletarian literature for *Kaebyŏk*. For more on *Kaebyŏk*, see Kim Kunsu, *Han'guk Chapchisa Yŏngu*, 153; and Robinson, *Cultural Nationalism in Colonial Korea*, 115.

5. *Chosŏn Chikwang* published its first issue on November 1, 1922, and was a magazine devoted to disseminating articles on "socialist theory, social movements of the time, and proletarian literature." After having withstood frequent confiscations, it was finally closed in 1930. Kim Kŭnsu, *Han'guk Chapchisa Yŏngu*, 125.

6. *Sin'gyedan* came out in 1932 and was edited and published by the same man who had edited *Chosŏn Chikwang*, Yu Jin-Hŭi. It closed in September 1933 after having published only eleven issues. Ibid., 151.

7. *Chosŏn Mundan* first appeared in 1924 and became one of the leading literary journals of the period. It published the proletarian literature of Yi Pukmyŏng and Pak Hwa-sŏng. For more on *Chosŏn Mundan*, see ibid., 126–37; and M. Shin, "Interior Landscapes," 427n90.

8. The nationalist newspapers the *Chosŏn Ilbo* and the *Tonga Ilbo* were the largest circulation dailies in the colony until they were both closed in 1940. Like the magazine *Kaebyŏk*, they were revived after liberation in 1945.

9. For an examination of the how the leadership structures of the labor movement were dominated by men in this period, see Yi Ok-ji, *Han'guk Yŏsŏng Nodongja Undongsa 1*, 39. For a discussion of how this was to the detriment of the labor movement as a whole, see Han'guk Yŏsŏng Yŏnguhoe, *Han'guk Yŏsŏngsa—Kŭndaepyon*, 241.

10. The phrase is Patricia Johnson's, who uses it in a different context: to describe the character Stephen Blackpool in Charles Dickens's *Hard Times*, who is physically and sexually dominated by his wife in a role reversal of conventional accounts of domestic violence in working-class homes. See Johnson, *Hidden Hands*, 149.

11. William Empson provides a succinct contemporary appraisal of this movement in his essay "The Proletarian Pastoral," published in 1934. See Empson, *Some Versions of Pastoral*.

12. Eagleton, *Rape of Clarissa;* Clark, "Politics of Seduction in English Popular Culture, 1748–1848," 47–70; and Denning, "Only a Mechanic's Daughter," in *Mechanic Accents*.

13. In suggesting the usefulness of the trope of seduction to analyze this literature, I am not attempting to valorize seduction in a way that reinforces an unambiguous distinction between seduction (nice) and rape (bad). Rather, I seek to explore the unresolvable ambiguities of seduction in relationships that thrive in circumstances of inequality and collaboration. For an excellent discussion of seduction in literature, see Rooney, "Criticism and the Subject of Sexual Violence."

14. The two story titles "Yŏgong" and "Yŏjikkong" are very similar and both can be translated as "Factory Girl." To distinguish between the two, I refer to them as "Factory Girl" and "The Textile Factory Girl," respectively. Yi Puk-myŏng was a member of KAPF. Yu Jin-o did not officially join KAPF and might best be described as a "fellow traveler" who supported its political line and wrote proletarian literature for the *Chosŏn Ilbo*, works that the South Korean state later banned. Both Yu Jin-o and Pak Yŏng-hŭi, another leading author of proletarian literature, in 1939 joined the Association of Chosŏn Literary Figures, a literary group instituted by the colonial state for the production of literature favorable to the Japanese imperial state.

15. Yi Puk-myŏng, "Yŏgong," 220. "Red love" *(pŭlkun yŏnae)* or "love between comrades" *(tongji yŏnae)* were two phrases coined in the 1920s to describe radical socialist couples who openly flouted feudal traditions concerning marriage and campaigned against repressive social conditions.

16. Yu Jin-o, "Yŏjikkong," 271–310.

17. *Chŏnjung* is the Korean pronunciation of a Japanese name, and the character is identified as Japanese in the text. I am indebted to Jin-Kyung Lee for pointing this out.

18. Yu Jin-o, "Yŏjikkong," 297.

19. Ibid., 310.

20. Korean migration to Kando had begun in the early nineteenth century as large numbers of farmers, ruined by bad harvests, migrated to Manchu-

ria in search of a better life. Following Japan's annexation of Korea in 1910, political exiles joined the economic migrants and Kando became a regional headquarters of the anti-Japanese independence movement. See Eckert et al., *Korea Old and New*, 185, 243, 273; and Yi Yi-hwa, *Han'guk Kŭnhyŏndaesa Sajon*, 90–93.

21. The following information about Kang Kyŏng-ae's life is taken from Yi Sang-gyŏng, *Kang Kyŏng-ae Chŏnjip*, 815–20; Kang Yi-su, "Singminjiha Yŏsŏng Munje-wa Kang Kyŏng-ae-ŭi *In'gan Munje*"; Sŏ Ŭn-ju, "Kang Kyŏng-ae"; Kim Yang-sŏn, "Kang Kyŏng-ae"; and Yi Hŭi-j'un, "Kang Kyŏng-ae Sosŏl Yŏngu."

22. These new missionary schools had their own class distinctions. Thus, one female politician informed the anthropologist Chung-Hee Soh that neither she nor any other upper-class girl would have considered attending Ewha Girls School, the first modern school for females in Korea, "because its student body in the early days was composed of concubines and other low class girls." Quoted in Chung-Hee Soh, *Chosen Women in Korean Politics*, 33.

23. Kim Yang-sŏn, "Kang Kyŏng-ae," 348; Yi Sang-gyŏng, *Kang Kyŏng-ae Chŏnjip*, 815. These experiences are also mentioned in Kang Kyŏng-ae's writings, "Wŏlsakŭm" (School Fees) and "Wŏnkoryo Yi Baek Wŏn" (200 Won Manuscript).

24. Kang Kyŏng-ae, "Wŏlsakŭm" and "Wŏnkoryo Yi Baek Wŏn."

25. See *Tonga Ilbo* October 17 and 18, 1923.

26. Yi Sang-gyŏng, *Kang Kyŏng-ae Chŏnji*, 816.

27. See H. Kang, *Under the Black Umbrella*, chap. 8.

28. Kyeong-Hee Choi notes that Kang Kyŏng-ae was on the periphery of the KAPF circle; she was a sympathizer but never joined the organization. Kyeong-Hee Choi, "Impaired Body as Colonial Trope," 441n31. Yi Sang-gyŏng, a leading authority on Kang Kyŏng-ae's literature and the editor of her collected works, notes that Kang Kyŏng-ae, in a review she published in the *Chosŏn Ilbo* in 1929, signed herself "Kang Kyŏng-ae of the Changyŏn branch of Kŭnuhoe." Yi Sang-gyŏng, *Kang Kyŏng-ae Chŏnjip*, 705.

29. See Yi Sang-gyŏng, *Kang Kyŏng-ae Chŏnjip*, 4–5.

30. See Sŏ Ŭn-ju, "Kang Kyŏng-ae"; Kim Yang-sŏn, "Kang Kyŏng-ae"; and Yi Hŭi-j'un, "Kang Kyŏng-ae Sosŏl Yŏngu."

31. The characters' names evoke their class and class pretensions. Thus Sŏnbi, who longs to be educated, is given an elevated name meaning "scholar," and Ch'ŏtchae has a throwaway name meaning "first born," common for children in large peasant families. Kannani's name means "new born," an incongruous name for an adult woman but one that speaks of her origins in an illiterate peasant family.

32. Kang Yi-su, "Singminjiha Yŏsŏng Munje-wa Kang Kyŏng-ae-ŭi *In'gan Munje*," 338.

33. Patricia Johnson, in *Hidden Hands*, 12, uses this phrase to describe the plight of working-class women in industrial England.

34. I am grateful to Suk-young Kim for alerting me to the rich history of

the concept and cultural phenomenon of the "superfluous man" in Russian literature dating to Pushkin's *Eugene Onegin* and Lermontov's *A Hero of Our Time*. Here I am only engaging with a shorthand version of the term that was operative in proletarian literary circles in colonial Korea.

35. The proletarian author Kim P'al-bong likened colonial Korea to Russia under Alexander III (1881–94) and called for people to choose between being "a Turgenev" and a laborer. Yoon-shik Kim, "Phases of Development of Proletarian Literature in Korea," 31. The South Korean author Son Ch'angsŏp wrote a short story titled "The Superfluous Man" after the Korean War. Russian literature, and indeed most literature from Western countries, was introduced to colonial Korea through Japanese translations; enthusiasm in Japan for particular writers and their works greatly influenced Korean literary tastes and affected the availability of works in translation.

36. Kang Kyŏng-ae, *In'gan Munje*, 71.

37. As in the story of Ch'ŭnhyang, the beautiful daughter of a *kisaeng* who secretly marries an aristocratic young man and then endeavors to deserve him by going through numerous public trials of her virtue. Kang Kyŏng-ae tells us that she read "The Tale of Ch'ŭnhyang" when she was eight years old and thus began her passionate attachment to classical Korean literature. Kang Kyŏng-ae, *Chasŏ Sojŏn*, 788–89.

38. See Eckert, *Offspring of Empire*, on the education program for textile workers at Kyongbang, which Eckert describes as "extremely useful in helping persuade reluctant parents to turn their children over to the factory" (201).

39. Carby, *Reconstructing Womanhood*, 22.

40. Terry Eagleton's assessment of the bind that Samuel Richardson's heroine Clarissa is caught in is relevant here. As Eagleton notes, "Power reproduces itself by engendering in its victims a collusion which is the very condition of their survival." Eagleton, *Rape of Clarissa*, 82.

41. Kang Kyŏng-ae, *In'gan Munje*, 238.

42. Ibid., 248–49.

43. Ibid., 250.

44. Sŏnbi has the coarse, gnarled hands of a person who has performed manual labor since childhood. Her hands contrast markedly with her lovely appearance, and Sinch'ŏl in particular finds himself unable to reconcile her beauty with her aged and wrinkled hands.

45. Kang Kyŏng-ae, *In'gan Munje*, 251–53.

46. *Chosen* is the Japanese name for Korea employed during the colonial period. It is a Japanese-language variation of the word *Chosŏn*, the name of the ruling Chosŏn dynasty (1392–1910) when Japan annexed Korea.

47. Kim Dong-ch'un, "Shape of Korea's Modernity," 4. Kim Dong-ch'un is here referring only to socialist groups within colonial Korea. For an account of the socialist guerilla movement in Manchuria in the 1930s, see chap. 1 of Charles Armstrong's *North Korean Revolution, 1945–1950*.

48. Richard Mitchell, writing about the recanting of activists in Japan, has suggested that the large proportion of proletarian writers and other leftists

who recanted and "reembraced nationalism" following the mass arrests that began in 1928 is partly explained by the tactics of Japan's Justice Department, which sought to reeducate leftists to "love" the nation rather than punish them for having rejected it. Richard Mitchell, *Thought Control in Prewar Japan* (Ithaca, NY: Cornell University Press, 1976), 98–99. I am grateful to Heather Bowen-Struyk for this reference.

49. Yoon-shik Kim, "Phases of Development of Proletarian Literature in Korea," 33.

50. The P'yŏngyang Red Labor Union was a revolutionary underground union that was part of a wider communist labor movement, with branches in Wonsan, Haeju, Yongsan, Yŏsu, and Masan. It was directly influenced by the Comintern, and its platform included the following demands: an eight-hour day; equal pay for all workers regardless of age, gender, or ethnicity; the right to strike; the right to demonstrate; social insurance; and the right to form revolutionary organizations. Yi Yi-hwa, *Han'guk Kŭnhyŏndaesa Sajŏn*, 230. The colonial government came down hard on the Red Labor Unions and by 1933 had arrested about one thousand union activists. W. S. Park, *Colonial Industrialization and Labor in Korea*, 123.

51. This information about Kang Ju-ryŏng is taken from Ch'oi Min-ji, *Han'guk Yŏsŏng Undong Sosa* [A Short History of the Women's Movement in Korea], reproduced in *Yŏsŏng1* (1985): 288; and Yi Ok-ji, *Han'guk Yŏsŏng Nodongja Undongsa 1*, 50–51.

52. Together with the only other prominent female author of proletarian literature of the time, Pak Hwa-sŏng, Kang Kyŏng-ae was dubbed a "masculine writer" by literary critics. Quoted in Sŏ Ŭn-ju, "Kang Kyŏng-ae," 296. The association of proletarian literature with "masculine" writing points to a more essential equation of proletarian with male. It also assumes that the practice of writing about proletarian characters is unfeminine, even when the characters, and the author, are female.

53. Kang Kyŏng-ae discusses this point in "Songnyŏnsa" (New Year's Message), published in *Shin Kajong* in December 1933.

54. Rancière, *Nights of Labor*, 23.

55. Kang Kyŏng-ae, "Songnyŏnsa," *Shin Kajong* (December 1933), reproduced in Yi Sang-gyŏng, *Kang Kyŏng-ae Chŏnjip*, 746.

56. Guillory, *Cultural Capital*, 349n25.

CHAPTER 3

1. In fact, President Park Chung Hee (1961–79) was trained in the Japanese Kwantung Army, which he entered in 1940 before joining the Republic of Korea Army set up by the American military government after liberation in 1945.

2. Amsden, *Asia's Next Giant*, 18–20.

3. These figures are taken from ibid., 222.

4. Ibid., 252.

5. Park's official title was chairman of the Supreme Council for National Reconstruction before he became president with the 1963 elections.

6. A government recruitment poster from 1970 captures this nicely, with a smiling, pretty, and well-groomed blue-collar woman calling for other women to become model workers and "learn the real value of labor." S. Kim, *Class Struggle or Family Struggle*, 5–6.

7. N. Armstrong, *Desire and Domestic Fiction*, 254.

8. Song Hyo-sun, *Sŏulro Kanŭnkil*, 27.

9. Koo, *Korean Workers*, 35.

10. Kim Won, *Yŏgong 1970*, 195.

11. Ibid.

12. Ibid.

13. For an account of the *minjung* (people's) cultural movement in the 1970s and 1980s, especially in literature, and its relationship to political and social movements in wider society, see Choi Hyun-moo, "Contemporary Korean Literature."

14. A number of feminist historians acknowledge the women's labor movement of the 1970s as having reignited the feminist movement in South Korea. See Yi Sŏng-hŭi, "Hyŏndae-ŭi Yŏsŏng Undong," 403.

15. Johnson, *Hidden Hands*, 8.

16. These are Chang Nam-su's words, in *Ppaeatkin Ilt'ŏ*, 36.

17. Other works from this period include Yu Dong-wu's *Onŭ Dolmaengi-ŭi Oech'im* (The Cry of One Stone; 1984); Pak Yŏng-gŭn's edited collection of worker's writings, *Kongjang Oksang-e Olla* (To the Factory Roof; 1984); an edited collection by Kim Kyŏng-suk, *Kŭrŏna Urinŭn Ŏje-ŭi uriga Anida* (But We Are Not Yesterday's We; 1986); and Na Po-sun's edited collection, *Uridŭl Kajin kŏt Pirok Chŏgŏdo: Kŭllojadŭl-ŭi kŭlmoŭm 1* (Even though We Don't Have Much: Workers' Collected Writings, vol. 1; 1983). In addition, unpublished collections of stories and autobiographical writings also circulated in factories, night schools, and literary circles throughout the 1980s. Kim Pyŏng-ik discusses this process in "Recent Korean Labor Novels," 12–13.

18. *Chŏnnohyŏp: Chuyo P'allye*, 385. Chŏnnohyŏp was an illegal organization and the umbrella organization representing democratic or independent unions in South Korea would not be legalized until March 10, 1997.

19. Koo, "Work, Culture, and Consciousness of the Korean Working-Class," 53—76. Also see Koo, *Korean Workers*.

20. Chŏng Mi-suk, "70 Nyŏndae Yŏsŏng Nodong Undong-ŭi Hwalsŏng-e Kwanhan Kyŏnghŏm Segyejok Yŏngu."

21. Chŏng Hyŏn-baek, "Yŏsŏng Nodongja-ŭi Ŭisik kwa Nodong Segye," 116–62.

22. Chang Nam-su, *Ppaeatkin Ilt'ŏ*, 7.

23. Koo, *Korean Workers*, 33.

24. Amsden, *Asia's Next Giant*, 79.

25. South Korea as a "rich and strong nation" is a famous Park Chung Hee pronouncement.

26. Koo, *Korean Workers*, 35.

27. Hagen Koo cites an Economic Planning Board study that estimates that in 1985 women made up 88 percent of garment workers, 77 percent of textile workers, and 68 percent of electronics workers. Ibid., 36.

28. Woo, *Race to the Swift*, 126.

29. Ibid., 144.

30. According to Chŏng Mi-suk, in "70 Nyŏndae Yŏsŏng Nodong Undong-ŭi Hwalsŏng-e Kwanhan Kyŏnghŏm Segyejok Yŏngu," 37, the large number of young females employed in the textile and apparel industries—in 1963, 67.9 percent of workers were women; in 1971, 69.7 percent—led to these factories being nicknamed "women's industries" *(yŏsŏng sanŏp)*.

31. The fading away of working-class women from public culture in the years following the Korean War was no accident. As Chung-hee Soh writes, "A phase of transition and backlash in women's activism set in after the military coup. In 1961, the military government disbanded all voluntary associations as well as political parties. Two years later, when women's organizations were allowed to reform, they opted to be politically neutral. During the decade of the 1960s, the activities of women's organizations were limited to those concerning self-improvement and volunteer social work for national reconstruction. Women's liberation activities were not pursued." Soh, *Chosen Women in Korean Politics*, 84.

32. Yi Ok-ji, in *Han'guk Yŏsŏng Nodongja Undongsa 1*, 69, mentions three major disputes at textile factories in the 1950s in which women workers played important roles.

33. "To get an idea of how acutely the population of Seoul was increasing in this period, after 1959 a million person increase was recorded after 5 years (1964), 4 years (1968) and 2 years (1970)." Chŏng Sung-kyo, "Seoul Under Construction," 58–59.

34. S. Kim, *Class Struggle or Family Struggle*, 7.

35. Ibid., 103.

36. Ibid., 104.

37. Chŏng Hyŏn-baek, "Yŏsŏng Nodongja-ŭi Ŭisik kwa Nodong Segye," 121.

38. Amsden, *Asia's Next Giant*, 203.

39. Sŏk Chŏng-nam, *Kongjang-ŭi Pulbitt*, 11.

40. Shin Kyŏng-suk, *Oettan Bang*, 58.

41. Sŏk Chŏng-name, 13–16.

42. Chŏng Mi-suk, "70 Nyŏndae Yŏsŏng Nodong Undong-ŭi Hwalsŏng-e Kwanhan Kyŏnghŏm Segyejok Yŏngu," 54–60.

43. Chang Nam-su, *Ppaeatkin Ilt'ŏ*, 33–34.

44. See Cho Wha Soon, *Let the Weak Be Strong*, 109–13, on the working conditions of bus girls and the successful Inch'ŏn bus girls' strike of 1982.

45. Chang Nam-su, *Ppaeatkin Ilt'ŏ*, 46.

46. See also Hong Se-hwa, *Nanŭn Ppari-ŭi T'aeksi Unjŏnsa*.

47. I borrow the term "labor feminism" from Michael Denning's *Cultural Front*, 145.

48. Williams, *Country and the City*, 13.

49. Christian Conference of Asia's Urban Rural Mission, "Interview with Hai Tai Workers," in *From the Womb of Han*, 61.

50. Chŏng Mi-suk, "70 Nyŏndae Yŏsŏng Nodong Undong-ŭi Hwalsŏng-e Kwanhan Kyŏnghŏm Segyejok Yŏngu," 60.

51. Ibid.

52. Ibid., 61.

53. Chŏng Myŏng-ja, *Naega Alul K'aego Naon Sungan*, 91, quoted in Chŏng Mi-suk, "70 Nyŏndae Yŏsŏng Nodong Undong-ŭi Hwalsŏng-e Kwanhan Kyŏnghŏm Segyejok Yŏngu," 61.

54. See Shin Kyŏng-suk, *Oettan Bang*. Both Heejae and the girl in the sweet-wrapping factory seem to be attempting to ease the narrator's potential recoil from their disfigured hands by telling the story of their injuries with levity.

55. A *shida* is an apprentice in a garment factory. When Chŏn T'ae-il carried out his survey of the lives of garment factory workers in Seoul's Peace Market in 1970, he found that four thousand *shidas* between the ages of twelve and twenty-one worked in the market for an average of fourteen hours a day. They earned between 1,800 and 3,000 won a month, a salary that made it difficult to feed and clothe themselves adequately, especially in winter. See Chŏn T'ae-il, *Nae ChukUmŭl Hŏtdoei Malla*, 181–82.

56. Christian Conference of Asia's Urban Rural Mission, *From the Womb of Han*, 17. This extract is taken from an anonymous article, "The Human Beings Market," translated from the Japanese by Miyazawa Teruko and the American Friends Service Committee. The article is described as the "personal story of daily experiences . . . written in 1976 by a woman working as a sewer in the Peace Garment District in Seoul."

57. Chŏng Mi-suk, "70 Nyŏndae Yŏsŏng Nodong Undong-ŭi Hwalsŏng-e Kwanhan Kyŏnghŏm Segyejok Yŏngu," 61–62.

58. Chang Nam-su, *Ppaeatkin Ilt'ŏ*, 102.

59. Sŏk Chŏng-nam, quoted in Kim Won, *Yŏgong 1970*, 554.

60. Pak Sun-hui, quoted in Kim Won, *Yŏgong 1970*, 553.

61. Song Hyo-sun, *Sŏulro Kanŭnkil*, 116–17, 158.

62. Kim Won, *Yŏgong 1970*, 548.

63. Christian Conference of Asia's Urban Rural Mission, *From the Womb of Han*, 16.

64. See Yu Dong-wu, *Onŭ Dolmaengi-ŭi Oech'im*, 44.

65. For more on the term *kongsuni* and its association with earlier forms of address for slaves, see Koo, *Korean Workers*, 62. For an example of the use of *kongsuni* as a term of abuse, see chap. 4 of the present volume. Carter Eckert, in "The South Korea Bourgeoisie," 113–15, also discusses the connection between institutionalized slavery among the landowning class in the early twentieth century and before, and attitudes of South Korean capitalists toward their working-class employees.

66. Kim Kyŏng-suk, *Kŭrŏna Urinŭn Ŏje-ŭi uriga Anida,* 106, quoted in Koo, *Korean Workers,* 137.

67. Elaine Kim quotes this figure from a National Assembly report of the same year in "Men's Talk," 1104. Although in the 1990s the South Korean sex industry was globalized in new ways, as Eastern European, South East Asian, and Russian women entered the sex economy in South Korea as prostitutes, dancers, and hostesses, in the 1980s and 1970s prostitutes in South Korea were predominantly Korean.

68. E. Kim, "Men's Talk," 109n4. *Kisaeng* tourism is the peddling of Korea's traditional female entertainment business to foreign clients.

69. Johnson, *Hidden Hands,* 57.

70. For an account of the sexual violence against women workers during the Tongil dispute, see Sŏk Chŏng-nam, *Kongjang-ŭi Pulbitt,* 90–91. Sin Illyŏng describes the sexual violence perpetrated by the *kusadae,* or "save the company" corps, a collection of factory foremen, guards, and hired thugs or men recently released from prison, who were paid to break up strikes, in *Yŏsŏng, Nodong, Pop,* 294–97. It should be noted that the *kusadae* were a phenomenon of the 1930s and 1980s.

71. Lee Hyo-chae, "Industrialization and Women," 147.

72. Koo, *Korean Workers,* 48.

73. Patricia Johnson, in *Hidden Hands,* 10, makes this same point about the classification of women with child workers in Britain's 1847 legislation that restricted their working hours.

74. For an account of how the "father-son" relationship operated in male-dominated factories, see Ogle, *South Korea,* 49–51.

75. For an excellent account of the popular songs that expressed this estrangement from family and from the past, see Yi Yŏng-mi, "Taejung Kayo Sŏk-ŭi Padawa Ch'ŏldo," 96–114.

76. Cho Wha Soon, *Let the Weak Be Strong,* 55.

77. "In the political movement group or labor movement group, a woman's opinion is often ignored or marginalized by the men. No matter how hard she tries to be heard, she will be pushed away from the centre against her will and eventually be left out completely. This kind of thing happens quite often. I myself have had such experiences. And the wounds from these experiences are very deep." Cho Wha Soon, *Let the Weak Be Strong,* 136 passim.

78. Chang Nam-su, *Ppaeatkin Ilt'ŏ,* 27.

79. Ibid., 48.

80. For more on the symbolic power of factories in South Korean development, see Kim Eun-shil, "Making of the Modern Female Gender," 182.

81. See also Koo, *Korean Workers,* 63–64.

82. Christian Conference of Asia's Urban Rural Mission, *From the Womb of Han,* 68, reports that, following the economic slowdown in 1979, working hours went up in the textile and electronics industries.

83. Alice Amsden, in *Asia's Next Giant,* 205–6, makes the interesting point that South Korea's long workweek is a legacy of the work practices that Japan

introduced when it colonized Korea and that Japan itself during its early period of industrialization modeled its practice on the lengthy workweek of Germany and France.

84. Han'guk Yŏsŏng Nodongjahoe, *Han'guk Yŏsŏng Nodongja-ŭi Hyŏnjang,* 32. See also Koo, *Korean Workers,* 59.

85. See Chŏn T'ae-il, *Nae Chukumŭl Hŏtdoei Malla,* 97.

86. S. Kim, *Class Struggle or Family Struggle,* 159; Cho Wha Soon, *Let the Weak Be Strong,* 108–9. See also Song Hyo-sun, *Sŏulro Kanŭnkil,* 60; and Pak Yŏng-gŭn, *Kongjang Oksang-e Olla,* 128, 139.

87. Janelli and Yim, *Making Capitalism,* 225–28; Koo, *Korean Workers,* 64–66.

88. Koo, *Korean Workers,* 67.

89. S. Kim, in "Women Workers and the Labor Movement in South Korea," 228, writes, "They received neither the protected status promised to women by patriarchal families nor the freedom promised to women by their participation in the labor market."

90. "Confucian capitalism" is a widely contested term. Janelli and Yim discuss both the problems and the illuminations that arise when applying it to the ideologies practiced by Korean companies. They focus on how Confucian "moral claims" filtered into management strategies in the 1980s, where "the use of cultural constructions and the pursuit of material interests were brought into congruity." Janelli and Yim, *Making Capitalism,* 239.

91. Seung-kyung Kim also makes this point in *Class Struggle or Family Struggle,* 172.

92. Ibid., 5–6.

93. Perhaps the most common example of a sex-segregated activity was smoking. An activity that young women could not indulge in publicly unless they were willing to risk abuse, spitting, and perhaps a slap, it became a treasured secret vice for many, many women.

94. S. Kim, *Class Struggle or Family Struggle,* 6.

95. Cho Wha Soon, *Let the Weak Be Strong,* 51. For a discussion of the informal, insulting language used to address working-class women, see chap. 4 in the present volume.

96. Cho Wha Soon, *Let the Weak Be Strong,* 51.

97. Chŏng Mi-suk, "70 Nyŏndae Yŏsŏng Nodong Undong-ŭi Hwalsŏng-e Kwanhan Kyŏnghŏm Segyejok Yŏngu," 67.

98. Yi Ok-ji, *Han'guk Yŏsŏng Nodongja Undongsa 1,* 137.

99. Pak Myŏng-rim, "Han'guk Chŏnjaeng-ŭi Chaengjŏm," 209.

100. N. Lee, in "South Korea Student Movement," 32n85, reports that, before a partial lifting of the ban on contraband texts in 1983, the list of banned texts included E. H. Carr's *What Is History?,* possession of which could bring a minimum prison sentence of one year.

101. Cho Wha Soon, *Let the Weak Be Strong,* 124.

102. Ibid., 105.

103. Ibid., 55.

104. Christian Conference of Asia's Urban Rural Mission, *From the Womb of Han*, 53.

105. The KBS television station chose to concentrate on the Tongil unionists' story in its history of the 1970s in the documentary program *Yŏngsang Sillok Haebang 50 Nyŏn—1945–1995* (A True History of the 50 Years Since Liberation), which aired in 1995. George Ogle also concentrates on this union in his book *South Korea: Dissent within the Economic Miracle*.

106. The following account of the Tongil Labor Union is taken from Han'guk Kidokkyo Kyohoe Hyŏbuihoe, *Nodong Hyonjang kwa Chungon*, 369 passim; Cho Wha Soon, *Let the Weak Be Strong*; Sŏk Chŏng-nam, *Kongjang-ŭi Pulbitt*; Koo, *Korean Workers*; and Yi Ok-ji, *Han'guk Yŏsŏng Nodongja Undongsa 1*.

107. Cho Wha Soon and George Ogle, both Urban Industrial Mission members, stress the incipient feminist nature of the union election, but Seung-kyung Kim argues that the main aim was for ordinary workers to have a union that represented them and that the overturning of the gender hierarchy in Tongil was not a motive but a byproduct of that intention. See Cho Wha Soon, *Let the Weak Be Strong*; Ogle, *South Korea: Dissent within the Economic Miracle*; and S. Kim, *Class Struggle or Family Struggle* and "Women Workers and the Labor Movement in South Korea."

108. For a graphic account of the sexual violence that she and others experienced during the Tongil dispute, see Sŏk Chŏng-nam, *Kongjang-ŭi Pulbitt*, 90–91.

109. Johnson, *Hidden Hands*, 76.

110. Yi Sŏng-hŭi, "Hyŏndae-ŭi Yŏsŏng Undong," 404; Christian Conference of Asia's Urban Rural Mission, *From the Womb of Han*, 43–44.

111. Cho Wha Soon, *Let the Weak Be Strong*, 72, 104.

112. For a more detailed account of the activities of the national federated union body (the FKTU) and its corruption in the 1950s, and restructuring after the military coup of 1961, see Kim Yun-hwan, *Han'guk Nodong Undongsa*, chaps. 5 and 6; and H. Nam, *Building Ships, Building a Nation*, chap. 2.

113. Sŏk Chŏng-nam, *Kongjang-ŭi Pulbitt*, preface.

114. Christian Conference of Asia's Urban Rural Mission, "Members of the Y.H. Trading Company Branch of [the] National Textile Workers Union, August 14, 1979," in *From the Womb of Han*, 34. Also quoted in Han'guk Kidokkyo Kyohoe Hyŏbuihoe, *Nodong Hyonjang kwa Chungon*, 579.

115. See *Tonga Ilbo*, August 10 and 11, 1979; and *Hang'uk Ilbo*, August 12, 1979.

116. This episode, the death of Kim Kyŏng-suk, has also resonated in intervening years. In her novel *The Solitary Room*, Shin Kyŏng-suk discusses the death/killing of Kim Kyŏng-suk. There is also a documentary film and several nonfiction accounts, as well as investigations carried out by the Presidential Truth Commission on Suspicious Deaths.

117. In response to mounting domestic, economic, and national security concerns building from the late 1960s, the Park Chung Hee government in 1972 declared martial law and instituted the Yushin constitution. President

Park's *yushin* (translated as "revitalizing") constitutional changes allowed him to appoint one-third of the National Assembly and granted him almost complete executive power over all arenas of government, enabling him to muzzle any opposition. He also possessed extensive emergency powers, coupled with a very efficient and well-funded secret police. See Eckert et al., *Korea Old and New,* 359–75.

118. For an account of the three major vanguard revolutionary groups of the 1960s and 1970s, see Han'guk Yŏksa Yŏnguhoe, *Han'guk Hyŏndaesa 3,* 171–206.

119. S. Kim, *Class Struggle or Family Struggle,* 109.

120. Ibid., 16.

121. Marshall Berman uses this phrase, in a different context, to describe the 1905 demonstration of petitioners to the Russian tsar, who were shot down outside the Winter Palace. Berman, *All That Is Solid Melts into Air,* 251.

122. From Tongil pangjik pokjik t'ujaeng wiwonhoe, *Tongil Pangjik Nodongjohap Undongsa* [The History of the Tongil Textile Labor Union] (Seoul: Tolbekae, 1985), 73–74, quoted in Koo, *Korean Workers,* 82. Translation by Koo.

123. S. Kim, *Class Struggle or Family Struggle,* 171.

124. The charges were violating Presidential Emergency Decree no. 9, a decree that made it a criminal offense to criticize the president, and violating the Law on Meetings and Demonstrations. Cho Wha Soon, *Let the Weak Be Strong,* 97.

125. As in the following exchange: "The prosecutor asked me: Was not your . . . action for the purpose of class struggle? I answered 'I do not know. I am ignorant about such things and do not understand what is meant by class struggle. I have only tried to follow the words of the Bible, acting as a shepherd to find the one lost sheep, though this means leaving the ninety-nine others to do so.'" Cho Wha Soon, *Let the Weak Be Strong,* 99. And in her summing up: "In Luke's Gospel, some people came to Jesus and warned, 'Herod is trying to kill you! Please hurry and escape!' But instead of escaping, Jesus told them, 'Go and tell the fox that today and tomorrow, I will chase Satan away and heal the sick, then the third day I will finish my work. Today, tomorrow and the day after, I have to go my way. Could a prophet die anywhere else but in Jerusalem?' As a disciple of Jesus, I am doing this work with the same mind as Jesus.' I turned to the courtroom audience and shouted, 'Even though a crowd of devils like Herod tried to kill me, I will fight against them without fear of death. My friends the workers are the same. We will fight the devils of this land, not fearing death. The *han* [righteous anger] of all the oppressed, poor, and marginalized will turn into the sword—the dagger—of God's judgment and stab deep into the hearts of the devils'" (99–100).

CHAPTER 4

1. Johnson, *Hidden Hands,* 149.

2. President Park Chung Hee promised that the sacrifice was temporary

and would eventually be rewarded: "In order to increase our export volume, we have to produce good quality goods at lower prices than goods produced by other countries and this is impossible if wages are high. What will happen to us if export volume decreases because of high wages and high prices for goods? I want you to understand that both improvements in workers' lives and the growth of corporations depend on our national development, so I ask for your cooperation to take pride and responsibility for the establishment of the nation. I can assure you that the rapid growth of the economy due to the continuing expansion of exports will provide a prosperous future for our three million workers." Park Chung Hee, *Our Nation's Path*, 2–3.

3. An example of this literary censorship was the ban on the reprinting of Korean proletarian literature published in the 1920s and 1930s. This ban was lifted in 1987.

4. For a discussion of the decisive role of the middle class in the unraveling of the Chun Doo Hwan regime, see J. Choi, "Political Cleavages in South Korea."

5. Chungmoo Choi, "The Minjung Culture Movement and the Construction of Popular Culture in Korea."

6. Although I do not have definite sales figures, the books were all available in mainstream bookstores by the late 1980s.

7. Kwŏn Yŏng-min, *Haebang 40 Nyŏn-ŭi munhak, 1945–1985*, 318. Kwŏn Yŏng-min is here referring to a wider body of literature than I can consider in this book. See his *Haebang 40 Nyŏn-ŭi munhak* for more on "industrial literature."

8. Hall, *Canary Girls and Stockpots*, 39–40, quoted in J. Rose, *Intellectual Life of the British Working Classes*, 275.

9. Sŏk, *Kongjang-ŭi Pulbitt*, 54.

10. Ibid., 18.

11. Williams, *Problems in Materialism and Culture*, 221.

12. Quoted in Pawel, *A Poet Dying*, 126.

13. Tomalin, *Thomas Hardy*,17.

14. Williams, *Problems in Materialism and Culture*, 221.

15. Rancière, *Nights of Labor*, 64.

16. See, for example, Chang Nam-su's discussion of her friend Nam-ok's decision to go to Iran for work that would be "well paid but hellish" and her death there in a road accident one month later. Chang Nam-su, *Ppaeatkin Ilt'ŏ*, 56–58.

17. Sŏk, *Kongjang-ŭi Pulbitt*, 54.

18. In making this argument I am indebted to the research of Gi-Hyun Shin's "Politeness and Deference in Korean." For a rare example of equal address forms despite unequal class relations, see the section later in this chapter on class romance.

19. Sŏk, *Kongjang-ŭi Pulbitt*, 55.

20. Ibid.

21. The Tongil excrement incident occurred on February 21, 1978, when, at

the election of union delegates at Tongil, a group of men and two women, at the instigation of the company, smashed the election boxes and rubbed excrement onto the women workers who had come there to vote, while police and National Textile Union officials officiating at the election stood by. For more on this incident, see Cho Wha Soon, *Let the Weak Be Strong*, chap. 12; and Yi Ok-ji, *Han'guk Yŏsŏng Nodongja Undongsa 1*, 340–41.

22. Chang Nam-su, *Ppaeatkin Ilt'ŏ*, 65.

23. The Yushin period began in 1972 with President Park Chung Hee's constitutional changes that granted him extensive executive powers. See Eckert et al., *Korea Old and New*, 359–75.

24. S. Kim, *Class Struggle or Family Struggle*, 58.

25. See Song Hyo-sun, *Sŏulro Kanŭnkil*, 18, for a particularly heartbroken account of leaving school.

26. "When deciding how to rank us the fact of [a student's] poverty and destitution play an important part," she says. Chang Nam-su, *Ppaeatkin Ilt'ŏ*, 9.

27. Ibid., 12.

28. Quoted in Chŏng Mi-suk, "70 Nyŏndae Yŏsŏng Nodong Undong-ŭi Hwalsŏng-e Kwanhan Kyŏnghŏm Segyejok Yŏngu," 55.

29. Chang Nam-su, *Ppaeatkin Ilt'ŏ*, 13.

30. Ibid., 22.

31. Amsden, *Asia's Next Giant*, 223.

32. Ibid., 252.

33. Chang Nam-su, *Ppaeatkin Ilt'ŏ*, 17.

34. Chŏn Sang-suk, "Yŏsŏng, Kŭdŭl-ŭi Chikŏp," 234.

35. Ogle, *South Korea*, 107.

36. *Hakch'ul* literally means "to leave school," but it was used to describe those students who left their studies and careers to enter factories or rural areas to labor for the disadvantaged. The police term for student-workers was *wijang ch'wiŏp'ja*, which translates as "disguised workers."

37. Industrial mission work was modeled on the French worker-priest movement of Christian socialists, who in the 1940s and 1950s labored on the docks and in factory areas and took workers as the subject of their ministry.

38. Ogle, *South Korea*, 88.

39. Cho Wha Soon, *Let the Weak Be Strong*, 57.

40. Ibid., 58.

41. Chŏng Mi-suk, "70 Nyŏndae Yŏsŏng Nodong Undong-ŭi Hwalsŏng-e Kwanhan Kyŏnghŏm Segyejok Yŏngu," vii.

42. Chang Nam-su, *Ppaeatkin Ilt'ŏ*, 27.

43. Ibid., 28.

44. Rancière, *Nights of Labor*, ix.

45. Chang Nam-su, *Ppaeatkin Ilt'ŏ*, 56–57.

46. Chŏn T'ae-il, *Nae ChukUmŭl Hŏtdoei Malla*.

47. In Cho Yŏng-rae, *Chŏn T'ae-il Pyongchon*, 69–74.

48. See N. Lee, *Making of Minjung*, 1, 11, 13; and S. Kim, *Class Struggle or Family Struggle*, 133.

49. Ogle, *South Korea*, 99. George Ogle also cites a police record from 1985–86 that reports that 671 student-workers were arrested in that year.

50. Another example is the "fake student" incident at Seoul National University in 1984, when students discovered what they alleged was a police agent posing as a student. For more on this, see N. Lee, *Making of Minjung*, 34n106.

51. Yun Jŏng-mo, *Koppi*.

52. Rancière, *Nights of Labor*, 23.

53. S. Moon, "Economic Development and Gender Politics in South Korea," 271.

54. I borrow this turn of phrase from Michael Sprinker, who wrote, "Someone must ultimately pay the price for the privilege exercised by the ruling classes not to engage in productive labour." Sprinker, *Imaginary Relations*, 185.

55. Chŏng Hyŏn-baek, "Yŏsŏng Nodongja-ŭi Ŭisik kwa Nodong Segye," 126.

56. Wonpung Textiles would become famous as the center of one of the fiercest labor disputes of the 1970s. For more information, see Han'guk Kidokkyo Kyohoe Hyŏbuihoe, *Nodong Hyonjang kwa Chungon*, 403–8.

57. Yŏngdŭngp'o is an industrial suburb of Seoul that would become a center for labor protest in the 1980s.

58. Chang Nam-su, *Ppaeatkin Ilt'ŏ*, 35.

59. Ibid., 35.

60. Demian appears in several different guises in Hesse's novel, beginning with the schoolboy Demian but also appearing as a portrait painting, as Frau Eva, and as the author himself; and his image becomes symbolic of spiritual or self-awakening.

61. *Wŏlgan Taehwa* (Monthly Dialogue) is the magazine where Sŏk Chŏngnam made her literary debut.

62. Chang Nam-su, *Ppaeatkin Ilt'ŏ*, 35–36.

63. Ibid., 37.

64. Here Chang Nam-su uses the derogatory term *kongsuni* to refer to herself as something shabby and insignificant.

65. Chang Nam-su, *Ppaeatkin Ilt'ŏ*, 37.

66. Ibid., 38.

67. Ibid., 100.

68. Chŏng Hyŏn-baek, "Yŏsŏng Nodongja-ŭi Ŭisik kwa Nodong Segye," 126.

69. Cho, *Let the Weak Be Strong*, 76; See also Kim Won, *Yŏgong 1970*, 561.

70. Kim Seung-kyung, *Class Struggle or Family Struggle*, 73.

71. Pak Wansŏ, "T'i t'aimŭi monyŏ."

72. N. Lee, *Making of Minjung*, 286.

73. E. Kim, "Men's Talk," 142.

74. I borrow the term "class properties" from Patricia Johnson in her book *Hidden Hands*, 34.

75. Chang Nam-su, *Ppaeatkin Ilt'ŏ*, 42–43, translated by Seung-kyung Kim in *Class Struggle or Family Struggle*, 172.

76. Kim Won, "The True Character and Desires of Factory Girls," 45–46.

77. Chang Nam-su, *Ppaeatkin Ilt'ŏ*, 3.

CHAPTER 5

1. Shin Kyŏng-suk, *Oettan Bang*, 15.
2. See Yi Sang-gyŏng, *Han'guk Kŭndae Yŏsŏng Munhaksaron*, 283.
3. Shin Kyŏng-suk, *Oettan Bang*, 178.
4. The phrase "aesthetically healing powers" is from Marcus Wood's *Slavery, Empathy and Pornography*.
5. N. Armstrong, *How Novels Think*, 3.
6. Ibid.
7. Shin Kyŏng-suk, *Oettan Bang*, 145.
8. Ibid., 147–48.
9. Ibid., 150.
10. Ibid., 156.
11. Ibid., 197.
12. N. Armstrong, *How Novels Think*, 152.
13. Marcus, *Between Women, Friendship, Desire and Marriage in Victorian England*, 35.
14. Ibid., 45. The factory they get work at is the Tongnam Electronics Company. At this time, electronics companies were at the upper end of the scale of factories, higher than garment- or wig-manufacturing factories because they were perceived to require the latest technological skills and to employ advanced management practices.
15. Ibid., 34.
16. Ibid., 86.
17. Ibid., 87.
18. Ibid., 70.
19. Ibid., 202.
20. Shin Kyŏng-suk, *Oettan Bang*, 43.
21. Ibid., 47.
22. Ibid., 80–81.
23. Ibid., 21–22.
24. Ibid., 35–36.
25. The French translation of *Oettan Bang (La Chambre Solitaire)* won the Prix de l'Inaperçu in 2009. It has also been translated and published in German, Japanese, and Chinese.
26. Paik Nak-chung, "The Hidden and the Revealed in Oettan Bang" [in Korean], republished as accompanying commentary in Shin Kyŏng-suk, *Oettan Bang*, 425–53.
27. Y. Yang, "Nation in the Backyard."
28. Rancière, "Good Times or Pleasure at the Barricades," 50.
29. N. Armstrong, *How Novels Think*, 9.
30. C. Brontë, *Villette*, 382.
31. Shin Kyŏng-suk, *Oettan Bang*, 73.
32. Ibid., 136.
33. Ibid.

34. Gabriel Sylvian, interview with Shin Kyŏng-suk, *Azalea: Journal of Korean Literature and Culture* 2 (2008): 61.
35. Shin Kyŏng-suk, *Oettan Bang*, 189.
36. Ibid., 113.
37. Ibid., 115.
38. Ibid.
39. Ibid., 380.
40. Irigaray, *Why Different?*, 79.
41. N. Armstrong, *How Novels Think*, 14.
42. Rancière, *Flesh of Words*, 110.
43. Shin Kyŏng-suk, *Oettan Bang*, 25.
44. See Noritake, "Negotiating Space and Gender."

EPILOGUE

1. Han'guk Hyongsa Jongch'aek Yonguwon [Korea Institute of Criminology], "Research into the State of Rape Crimes, Korea Institute of Criminology" [in Korean], chap. 7 in *Poknyok Munhwa-wa Poknyoksŏng Pomjui* [South Korea's Culture of Violence and the Nature of Violent Crimes], 238 (1991). I am grateful to the staff of the Korean Sexual Violence Relief Center, especially Director Yi Eunsang, for allowing me access to their library and in particular for making this report available to me.
2. Insook Kwon, *Hana-ui Pyokul Nomoso*. For an excellent analysis of Insook Kwon's case, see Yi Sang-rok, "Siminul Song Pokhaenghanun Minjukuka, Taehan Minguk" [Republic of Korea: The State that Sexually Assaulted Its Citizens].

Selected Bibliography

JOURNALS AND NEWSPAPERS

Chosŏn Chikwang
Chosŏn Chungang Ilbo
Chosŏn Ilbo
Chosŏn Mundan
Hang'uk Ilbo
Kaebyŏk
Samcholli
Sin'gyedan
Shin Kajong
Shin Yŏsŏng
Tonga Ilbo
Wŏlgan Taehwa
Yŏsŏng

BOOKS AND ARTICLES

Abelmann, Nancy. *Echoes of the Past, Epics of Dissent.* Berkeley: University of California Press, 1996.

———. *The Melodrama of Mobility: Women, Talk and Class in Contemporary South Korea.* Honolulu: University of Hawai'i Press, 2003.

Amsden, Alice. *Asia's Next Giant: South Korea and Late Industrialization.* Oxford: Oxford University Press, 1989.

Armstrong, Charles. *The North Korean Revolution, 1945–1950.* Ithaca, NY: Cornell University Press, 2003.

Armstrong, Nancy. *Desire and Domestic Fiction: A Political History of the Novel.* New York: Oxford University Press, 1987.

———. *How Novels Think.* New York: Columbia University Press, 2006.

Barme, Geremie. *In the Red: On Contemporary Chinese Culture.* New York: Columbia University Press, 1999.

Benjamin, Walter. *The Arcades Project.* Cambridge, MA: Belknap Press, 1999.
Berman, Marshall. *All That Is Solid Melts into Air: The Experience of Modernity.* London: Verso, 1983.
Bourdieu, Pierre. *Acts of Resistance.* Cambridge: Polity Press, 1998.
———. "Postscript." In *The Weight of the World: Social Suffering in Contemporary Society,* ed. Pierre Bourdieu et al., 627–29. Stanford, CA: Stanford University Press, 1999.
Bowen-Struyk, Heather. "Rethinking Japanese Proletarian Literature." PhD dissertation, University of Michigan, 2001.
Brontë, Charlotte. *Villette.* London: Penguin Classics, 1985.
Brontë, Emily. *Wuthering Heights.* London: Penguin Classics, 2003.
Butler, Judith. "Gender as Performance: An Interview with Judith Butler." *Radical Philosophy* 67 (Summer 1994): 32–39.
———. *Gender Trouble: Feminism and the Subversion of Identity.* New York: Routledge, 1999.
Butler, Judith, and Joan Scott, eds. *Feminists Theorize the Political.* New York: Routledge, 1991.
Carby, Hazel. *Reconstructing Womanhood.* New York: Oxford University Press, 1987.
Ch'ae Manshik. "P'alyŏkan Mom" [Sold Body]. *Shin Kajong* (August 1933): 93–97. Republished in *Ch'ae Manshik Chŏnjip* [Collected Works of Ch'ae Manshik], 39–46. Seoul: Ch'angjaksa, 1987–89.
Chakrabarty, Dipesh. *Rethinking Working-Class History: Bengal, 1890–1940.* Princeton, NJ: Princeton University Press, 1989.
Chang Nam-su. *Ppaeatkin Ilt'ŏ* [The Lost Workplace]. Seoul: Ch'angjakkwa Bip'yŏngsa, 1984.
Chang Rin. "Nodong Puin-ŭi Chojikhwarul" [On the Organization of Women Workers]. *Kŭnu* [Rose of Sharon] (1929): 33–34.
Cho Se-hui. *The Dwarf.* Honolulu: University of Hawai'i Press, 2006.
Cho Wha Soon. *Let the Weak Be Strong: A Woman's Struggle for Justice.* Bloomington, IN: Meyer-Stone Books, 1988.
Cho Yŏng-rae. *Chŏn T'ae-il Pyongchon* [A Critical Biography of Chŏn T'ae-il]. Seoul: Tolbekae, 1983.
Choi, Chungmoo. "The Minjung Culture Movement and the Construction of Popular Culture in Korea." In *South Korea's Minjung Movement: The Culture and Politics of Dissidence,* ed. Kenneth Wells, 105–18. Honolulu: University of Hawai'i Press, 1995.
Choi, Hyun-moo. "Contemporary Korean Literature: From Victimization to Minjung Nationalism." In *South Korea's Minjung Movement: The Culture and Politics of Dissidence,* ed. Kenneth Wells, 167–78. Honolulu: University of Hawai'i Press, 1995.
Choi, Jang Jip. "Political Cleavages in South Korea." In *State and Society in Contemporary Korea,* ed. Hagen Koo, 13–50. Ithaca, NY: Cornell University Press, 1993.

Choi, Kyeong-Hee. "Impaired Body as Colonial Trope: Kang Kyŏng-ae's 'Underground Village'." *Public Culture* 13, no. 3 (2001): 431–58.
Ch'ŏn Chŏng-hwan. *Kŭndae-ŭi Ch'aek Ilgi: Tokcha-ŭi T'ansaeng kwa Hang'uk Kŭndae Munhak* [Reading Modern Books: Modern Korean Literature and the Birth of Readers]. Seoul: P'urŭn Yŏksa, 2003.
Chŏn Sang-suk. "Yŏsŏng, Kŭdŭl-ŭi Chikŏp" [Women and Their Careers]. In *Urinŭn Chinan Baengnyŏn Dongan Ŏt'ŏke Salassŭlkka? 2* [How Did We Live over the Last 100 Years?, vol. 2], 225–39. Seoul: Yŏksa Bip'yŏngsa, 1998.
Chŏn T'ae-il. *Nae Chukumŭl Hŏtdoei Malla* [Do Not Waste My Life]. Seoul: Tolbekae, 1988.
Chŏng Hae-Un. "Pongkŏn Ch'eje-ŭi Tongyo-wa Yŏsŏng Sŏngjang" [Disturbances in the Feudal Order and the Rise of Women]. In *Uri Yŏsŏng-ŭi Yŏksa* [A History of Korean Women], by Han'guk Yŏsŏng Yŏnguso [Institute for Research on Korean Women], 225–50. Seoul: Ch'ŏngnyŏnsa, 1999,.
Chŏng Hyŏn-baek. "Yŏsŏng Nodongja-ŭi Ŭisik kwa Nodong Segye: Nodongja Suki Punsŏkŭl Chunsimŭro" [Women Workers' Consciousness and the World of Work: Analyzing Workers' Writings]. *Yŏsŏng* [Women]1 (1985): 116–62.
Chŏng Mi-suk. "70 Nyŏndae Yŏsŏng Nodong Undong-ŭi Hwalsŏng-e Kwanhan Kyŏnghŏm Segyejok Yŏngu" [A Study on the Women's Labor Movement in the 1970s]. Master's thesis, Ewha Women's University, 1993.
Chŏng Myŏng-ja. *Naega Alul K'aego Naon Sungan* [The Moment I Split Open My Shell]. Seoul: Kongdongch'e, 1989.
Chŏng Sung-kyo. "Seoul Under Construction." In *Urinŭn Chinan Baengnyŏn Dongan Ŏt'ŏke Salassŭlkka? 2* [How Did We Live over the Last 100 Years?, vol. 2], 58–62. Seoul: Yŏksa Bip'yŏngsa, 1998.
Chŏnnohyŏp: Chuyo P'allye [Leading Case Studies in the History of Chŏnnohyŏp], vol. 2. Seoul: *Chŏnnohyŏ*p, 1998.
Chow, Rey. *Woman and Chinese Modernity: The Politics of Reading between West and East.* Minneapolis: University of Minnesota Press, 1991.
Clark, Anna. "The Politics of Seduction in English Popular Culture, 1748–1848." In *The Progress of Romance: The Politics of Popular Fiction,* ed. Jean Radford, 47–70. London: Routledge and Kegan Paul, 1986.
———. *The Struggle for the Breeches: Gender and the Making of the British Working Class.* Berkeley: University of California Press, 1995.
———. *Women's Silence, Men's Violence: Sexual Assault in England, 1770–1845.* London: Pandora Press, 1987.
Christian Conference of Asia's Urban Rural Mission, ed. *From the Womb of Han: Stories of Korean Women Workers.* Hong Kong: Christian Conference of Asia's Urban Rural Mission, 1982.
Coetzee, J.M. *Giving Offense: Essays on Censorship.* Chicago: University of Chicago Press, 1996.
Crook, Stephen. *Modernist Radicalism and Its Aftermath.* London: Routledge, 1991.

Cumings, Bruce. *The Origins of the Korean War: Liberation and the Emergence of Separate Regimes.* Princeton, NJ: Princeton University Press, 1981.
Damousi, Joy. *Women Come Rally.* Melbourne: Oxford University Press, 1994.
Denning, Michael. *The Cultural Front.* London: Verso, 1997.
———. *Mechanic Accents: Dime Novels and Working-Class Culture in America.* London: Verso, 1998.
Deyo, Frederic. *Beneath the Miracle: Labor Subordination in the New Asian Industrialism.* Berkeley: University of California Press, 1989.
Douglas, Mary. *How Institutions Think.* Syracuse, NY: Syracuse University Press, 1986.
Eagleton, Terry. *Heathcliff and the Great Hunger: Studies in Irish Culture.* London: Verso, 1996.
——— *Rape of Clarissa*, London: Basil Blackwell, 1982.
Eckert, Carter. *Offspring of Empire: The Koch'ang Kims and the Colonial Origins of Korean Capitalism.* Seattle: University of Washington Press, 1991.
———. "The South Korea Bourgeoisie: A Class in Search of Hegemony." In *State and Society in Contemporary Korea*, ed. Hagen Koo, 95–130. Ithaca, NY: Cornell University Press, 1993.
———"Total War, Industrialization, and Social Change in Late Colonial Korea." In *The Japanese Wartime Empire, 1931–1945*, ed. Peter Duus, Ramon Myers, and Mark Peattie, 3–39. Princeton, NJ: Princeton University Press, 1996.
Eckert, Carter, Ki-baik Lee, Michael Robinson, Edward Wagner, and Young Ick Lew. *Korea Old and New.* Seoul: Ilchokak, 1991.
Eliot, George. *Mill on the Floss.* New York: Harper and Row, 1965.
———. *Romola.* London: Penguin, 1980.
Empson, William. *Some Versions of Pastoral.* New York: New Directions, 1974.
Faison, Elyssa. *Managing Women: Disciplining Labor in Modern Japan.* Berkeley: University of California Press, 2007.
Gaskell, Elizabeth. *The Life of Charlotte Brontë.* Oxford: Oxford University Press, 1966.
———. *Mary Barton.* Oxford: Oxford University Press, 1987.
Guillory, John. *Cultural Capital: The Problem of Literary Canon Formation.* Chicago: Chicago University Press, 1993.
Hall, Edith. *Canary Girls and Stockpots.* London: Luton, 1977.
Han'guk Hyongsa Jongch'aek Yonguwon [Korea Institute of Criminology]. "Research into the State of Rape Crimes, Korea Institute of Criminology." In *Han'guk-ui Poknyok Munhwa-wa Poknyoksong Pomjui* [South Korea's Culture of Violence and the Nature of Violent Crimes], chap. 7. Seoul: Korean Institute of Criminology, 1991.
Han'guk Kidokkyo Kyohoe Hyŏbuihoe [National Council of Churches in Korea]. *Nodong Hyonjang kwa Chungon* [The Scene and Testimony of Labor]. Seoul: P'ulbitt, 1984.
Han'guk Nodong Chohap Ch'ongyŏnmaeng [Federation of Korean Trade Unions]. *Han'guk Nodong Chohap Undongsa* [History of the Korean Labor Union Movement]. Seoul: Koryŏ Sŏjŏk, 1979.

Han'guk Yŏksa Yŏnguhoe [Korean History Research Association]. *Han'guk Hyŏndaesa 3* [Contemporary Korean History, vol. 3]. Seoul: P'ulbitt, 1996.

Han'guk Yŏsŏng Nodongjahoe [Korean Women Workers' Association]. *Han'guk Yŏsŏng Nodongja-ŭi Hyŏnjang* [The Scene of Korean Women Workers]. Seoul: Paeksan Sŏdang, 1987.

Han'guk Yŏsŏng Yŏnguhoe [Association for Research on Korean Women]. *Han'guk Yŏsŏngsa—Kŭndaepyon* [A History of Korean Women—The Modern Period]. Seoul: P'ulbitt, 1992.

Henderson, Gregory. *Korea: The Politics of the Vortex*. Cambridge, MA: Harvard University Press, 1968.

Hong Se-hwa. *Nanŭn Ppari-ŭi T'aeksi Unjŏnsa* [I Was a Parisian Taxi Driver]. Seoul: Ch'angbi, 1995.

Insook Kwon. *Hana-ui Pyokul Nomoso* [Overcoming One Barrier]. Seoul: Korum, 1989.

Irigaray, Luce. *Why Different? A Culture of Two Subjects*. New York: Semiotext(e), 2000.

Jameson, Frederic. *Late Marxism: Adorno; or, The Persistence of the Dialectic*. London: Verso, 1996.

———. "Magical Narratives: Romance as Genre." *New Literary History* 7 (1975): 135–63.

Janelli, Roger, and Dawnhee Yim. *Making Capitalism: The Social and Cultural Construction of a South Korean Conglomerate*. Stanford, CA: Stanford University Press, 1993.

Johnson, Patricia. *Hidden Hands: Working-Class Women and Victorian Social-Problem Fiction*. Athens: Ohio University Press, 2001.

Joyce, Patrick, ed. *Class*. Oxford: Oxford University Press, 1995.

Kang, Hildi. *Under the Black Umbrella: Voices from Colonial Korea, 1910–1945*. Ithaca, NY: Cornell University Press, 2001.

Kang Kyŏng-ae. "Chakja-ŭi Mal." *Tonga Ilbo*, July 27, 1934.

———. "Chasŏ Sojŏn" [An Autobiographical Tale]. In *Kang Kyŏng-ae Chŏnjip* [Kang Kyŏng-ae's Complete Works], ed. Yi Sang-gyŏng, 778–89. Seoul: Somyŏng Ch'ulp'an, 1999.

———. *In'gan Munje* [The Human Predicament]. Seoul: Sodam Ch'ulp'ansa, 1996.

———. "Wŏlsakŭm" [School Fees]. In *Kang Kyŏng-ae Chŏnjip* [Kang Kyŏng-ae's Complete Works], ed. Yi Sang-gyŏng, 441–43. Seoul: Somyŏng Ch'ulp'an, 1999.

———. "Wŏnkoryo Yi Baek Wŏn" [200 Won Manuscript]. In *Kang Kyŏng-ae Chŏnjip* [Kang Kyŏng-ae's Complete Works], ed. Yi Sang-gyŏng, 559–67. Seoul: Somyŏng Ch'ulp'an, 1999.

Kang Yi-su. "1930 Nyŏndae Myŏnbang Taekiŏp Yŏsŏng Nodongja-ŭi Sangt'ae-e Taehan Yŏngu" [Research into the Situation of Women Workers in the Large Cotton Spinning Factories in the 1930s]. PhD dissertation, Ewha Women's University, 1992.

———. "Singminjiha Yŏsŏng Munje-wa Kang Kyŏng-ae-ŭi *In'gan Munje*"

[Kang Kyŏng-ae's *The Human Predicament* and the Situation of Women under Japanese Colonialism]. *Yŏksa Pib'yŏng* [History and Criticism] (Autumn 1993): 335–48.

Kawashima, Ken. *The Proletarian Gamble: Korean Workers in Interwar Japan.* Durham, NC: Duke University Press, 2009.

Kim, Elaine. "Men's Talk." In *Dangerous Women: Gender and Korean Nationalism,* ed. Elaine Kim and Chungmoo Choi, 67–117. London: Routledge, 1998.

Kim, Janice. *To Live to Work.* Stanford, CA: Stanford University Press, 2009.

Kim, Seung-kyung. *Class Struggle or Family Struggle: The Lives of Women Factory Workers in South Korea.* Cambridge: Cambridge University Press, 1997.

———. "Women Workers and the Labor Movement in South Korea." In *Anthropology and the Global Factory,* ed. Frances Rothstein and Michael Blim, 220–37. New York: Bergin and Garvey, 1992.

Kim, Uchang. "The Agony of Cultural Construction." In *State and Society in Contemporary Korea,* ed. Hagen Koo, 163–95. Ithaca, NY: Cornell University Press, 1993.

Kim, Yoon-shik. "Phases of Development of Proletarian Literature in Korea." *Korea Journal* 27, no. 1 (January 1987): 31–45.

Kim, Yung-Hee. "A Critique on Traditional Korean Family Institutions: Kim Wŏnju's 'Death of a Girl'." *Korean Studies* 23 (1999): 24–42.

Kim Dae-hwan. "Kŭndaejŏk Imgŭm Nodong-ŭi Hyŏngsŏng Kwajŏng" [Formation and Development of Modern Waged Work]. In *Han'guk Nodong Munje-ŭi Kujo* [The Structure of Labor Problems in Korea], 59–95. Seoul: Kwangminsa, 1978.

Kim Dong-ch'un. "The Shape of Korea's Modernity." Unpublished paper, 1996.

Kim Eun-shil. "The Making of the Modern Female Gender: The Politics of Gender in Reproductive Practices in Korea." PhD dissertation, University of California, 1993.

Kim Kunsu. *Han'guk Chapchisa Yŏngu* [Research on the History of Magazines in Korea]. Seoul: Han'gukhak Yŏnguso, 1992.

Kim Kyŏng-il. *Ilcheha Nodong Undongsa* [A History of the Labor Movement under Japanese Colonialism]. Seoul: Ch'angjakkwa Bip'yŏngsa, 1992.

———. *Yŏsong-ui Kundae, Kundae-ui Yŏsong* [Women's Modernity and Modernity's Women]. Seoul: Purun Yoksa, 2004.

Kim Kyŏng-suk. *Kŭrŏna Urinŭn Ŏje-ŭi uriga Anida* [But We Are Not Yesterday's We]. Seoul: Tolbegae, 1986.

Kim Pyŏng-ik. "Recent Korean Labor Novels: 'Labor' Literature vs. Labor 'Literature'." *Korea Journal* 29, no.3 (March 1989): 12–22.

Kim Won. "The True Character and Desires of Factory Girls: A Critical Study of 1970s 'Factory Girl Discourse'." *Sociology Research* 12, no. 1 (2004): 44–80.

———. *Yŏgong 1970: Ku-nyodul-ui Pan Yoksa* [1970 Factory Girls: A Counterhistory]. Seoul: Imagine Press, 2006.

Kim Yang-sŏn. "Kang Kyŏng-ae—Kando Ch'aehom kwa Chisikin Yŏsŏng-ŭi Chagipansŏng" [Kang Kyŏng-ae—Her Kando Experience and the Self-Reflections of a Female Intellectual]. *Yŏksa Pib'yŏng* [History and Criticism] (Summer 1996): 346–63.

Kim Yun-hwan. *Han'guk Nodong Undongsa* [The History of the Korean Labor Movement]. Seoul: Ch'ŏngsa, 1982.

Koo, Hagen. "From Farm to Factory: Proletarianization in Korea." *American Sociological Review* 55 (October 1990): 669–81.

———. *Korean Workers: The Culture and Politics of Class Formation.* Ithaca, NY: Cornell University Press, 2001.

———. "Work, Culture, and Consciousness of the Korean Working-Class." In *Putting Class in Its Place,* ed. Elizabeth Perry, 53–76. Berkeley: Institute of East Asian Studies, University of California, 1996.

Koven, Seth. *Slumming: Sexual and Social Politics in Victorian London.* Princeton, NJ: Princeton University Press, 2004.

Kwŏn Yŏng-min. *Haebang 40 Nyŏn-ŭi munhak, 1945–1985* [Korean Literature in the 40 Years Since Liberation, 1945–1985]. Seoul: Minumsa, 1985.

Lang, Amy Schrager. *The Syntax of Class: Writing Inequality in Nineteenth Century America.* Princeton: Princeton University Press, 2003.

Lang, Miriam. "San Mao and the Known World." PhD dissertation, Australian National University, 1999.

Lee, Hermione. *Virginia Woolf.* London: Vintage, 1997.

Lee, Hoon K. *Land Utilization and Rural Economy in Korea.* New York: Greenwood Press, 1936.

Lee, Ji-Eun. "New Women to New Housewives: Changing Discourses in Sinyŏsŏng, 1923–34," *U.S.-Japan Women's Journal,* no. 40 (2011): 90–121.

Lee, Namhee. *The Making of Minjung: Democracy and the Politics of Representation in South Korea.* Ithaca, NY: Cornell University Press, 2007.

———. "The South Korea Student Movement: Undongkwon as Counter-Public Sphere." In *Korean Society: Civil Society, Democracy and the State,* ed. Charles Armstrong, 95–120. London: Routledge, 2002.

Lee Hyo-chae. "Ilcheha-ŭi Yŏsŏng Nodong Munje" [The Situation of Women Workers in the Colonial Period]. In *Han'guk Nodong Munje-ŭi Kujo* [The Structure of Labor Problems in Korea], 131–79. Seoul: Kwangminsa, 1978.

———. "Industrialization and Women: The Social Background of Cho Wha Soon's Ministry." In *Let the Weak Be Strong,* ed. Cho Wha Soon, 146–50. Bloomington, IN: Meyer-Stone Books, 1988.

Mackie, Vera. *Creating Socialist Women in Japan: Gender, Labour and Activism, 1900–1937.* Cambridge: Cambridge University Press, 1997.

MacKinnon, Catherine. *Sexual Harassment of Working Women: A Case of Sex Discrimination.* New Haven, CT: Yale University Press, 1979.

Marcus, Sharon. *Between Women, Friendship, Desire and Marriage in Victorian England.* Princeton, NJ: Princeton University Press, 2007.

———. "Fighting Bodies, Fighting Words: A Theory and Politics of Rape Prevention." In *Feminists Theorize the Political*, ed. Judith Butler and Joan Scott, 385–403. New York: Routledge, 1991.

Milner, Andrew. *Cultural Materialism*. Carlton: Melbourne University Press, 1993.

Moon, Katherine. *Sex among Allies: Military Prostitution in US-Korea Relations, 1971–76*. New York: Columbia University Press, 1999.

Moon, Seungsook. "Economic Development and Gender Politics in South Korea: 1963–1992." PhD dissertation, Brandeis University, 1994.

Mun Kyŏng-ran. "Migun Chŏnggi Han'guk Yŏsŏng Undong-e Kwanhan Yŏngu" [Research into Korean Women's Movements under the American Military Government]. Master's thesis, Ewha Women's University, 1989.

Myers, Brian. *Han Sorya and North Korean Literature: The Failure of Socialist Realism in the DPRK*. Ithaca, NY: Cornell University Press, 1994.

Na Po-sun, ed. *Uridŭl Kajin kŏt Pirok Chŏgŏdo: Kŭllojadŭl-ŭi kŭlmoŭm 1* [Even though We Don't Have Much: Workers' Collected Writings, vol. 1]. Seoul: Tolbekae, 1983.

Nam, Hwasook. *Building Ships, Building a Nation: Korea's Democratic Unionism under Park Chung Hee*. Seattle: University of Washington Press, 2009.

Nam, Jeong-lim. "Gender Politics in the Korean Transition to Democracy." *Korean Studies* 24 (2000): 94–112.

Noritake, Ayami. "Negotiating Space and Gender: Female Street Entrepreneurs in Seoul." *Intersections: Gender and Sexuality in Asia and the Pacific* 17 (July 2008). http://intersections.anu.edu.au/issue17/noritake.htm.

Ogle, George. *South Korea: Dissent within the Economic Miracle*. London: Zed Books, 1990.

Pak Myŏng-rim. "Han'guk Chŏnjaeng-ŭi Chaengjŏm" [Issues in the History of the Korean War]. In *Haebang Chŏnhusa-ŭi Insik 6* [Understanding the History of the Liberation Period, vol. 6], 163–213. Seoul: Han'gilsa, 1989.

Pak Yŏng-gŭn, ed. *Kongjang Oksang-e Olla* [To the Factory Roof]. Seoul: P'ulbitt, 1984.

Pak Wansŏ. "T'i t'aimŭi monyŏ" [Mother and Daughter at Tea Time]. *Ch'angjak kwa Pip'yŏng* 21:2 (1992): 144–60.

Park, Won Soon. *Colonial Industrialization and Labor in Korea: The Onoda Cement Factory*. Cambridge, MA: Harvard University Asia Center, 1999.

Park Chung Hee. *Our Nation's Path*. Seoul: Hollym, 1970.

Pawel, Ernst. *The Poet Dying: Heinrich Heine's Last Years in Paris*. New York: Farrar, Straus and Giroux, 1995.

Pine, Adrienne. *Working Hard, Drinking Hard: On Violence and Survival in Honduras*. Berkeley: University of California Press, 2009.

Rabinowitz, Paula. *Labor and Desire: Women's Revolutionary Fiction in Depression America*. Chapel Hill: University of North Carolina Press, 1991.

Rancière, Jacques. *The Flesh of Words*. Stanford, CA: Stanford University Press, 2004.

———. "Good Times or Pleasure at the Barricades." In *Voices of the People:*

The Social Life of "La Sociale" at the End of the Second Empire, ed. Adrian Rifkin and Roger Thomas, 45–95. London: Routledge and Kegan Paul, 1998.

———. *The Nights of Labor: The Workers' Dream in Nineteenth-Century France.* Philadelphia: Temple University Press, 1989.

———. *The Philosopher and His Poor.* Durham: Duke University Press, 2003.

Robinson, Michael Edson. *Cultural Nationalism in Colonial Korea, 1920–1925.* Seattle: University of Washington Press, 1988.

Roh Jiseung. "The Pleasure of Lower-Class Women in 'Youngja's Best Days': A Cultural History of Class and Gender." *Han'guk Hyondae Munhak Yongu* [Reading Contemporary Korean Literature] 24, no. 4 (2008): 413–44.

Rooney, Ellen. "Criticism and the Subject of Sexual Violence." *Modern Language Notes* 98, no. 5 (December 1983): 1269–78.

Rose, Jonathan. *The Intellectual Life of the British Working Classes.* New Haven, CT: Yale University Press, 2001

Rose, Sonya. *Limited Livelihoods: Gender and Class in Nineteenth-Century England.* Berkeley: University of California Press, 1992.

Rowbotham, Sheila. *Beyond the Fragments: Feminism and the Making of Socialism.* London: Merlin Press, 1979.

Said, Edward. "Opponents, Audiences, Constituencies and Community." In *The Anti-Aesthetic: Essays on Postmodern Culture,* ed. Hal Foster, 135–59. Seattle: Bay Press.

Scalapino, Robert, and Chong-sik Lee. *Communism in Korea.* Vol. 1. Berkeley: University of California Press, 1972.

Scott, Joan Wallach. *Gender and the Politics of History.* New York: Columbia University Press, 1988.

Shiach, Morag. *Modernism, Labour, and Selfhood in British Literature and Culture, 1890–1930.* Cambridge: Cambridge University Press, 2004.

Shin, Gi-Hyun. "Politeness and Deference in Korean: A Case Study of Pragmatic Dynamics." PhD dissertation, Monash University, 1999.

Shin, Michael. "Interior Landscapes: Yi Kwangsu's 'The Heartless' and the Origins of Modern Literature." In *Colonial Modernity in Korea,* ed. Gi-Wook Shin and Michael Robinson, 248–87. Cambridge, MA: Harvard University Press, 1999.

Shin Kyŏng-suk. *Oettan Bang* [The Solitary Room]. Seoul: Munhak Tongnae, 1995.

Silverberg, Miriam. "The Modern Girls as Militant." In *Recreating Japanese Women, 1600–1945,* ed. Gail Bernstein, 239–66. Berkeley: University of California Press, 1991.

Sin Illyŏng. *Yŏsŏng, Nodong, Pŏp* [Women, Labor, Law]. Seoul: P'ulbitt, 1988.

Sin Yŏng-suk. "Ilche Singminjiha-ŭi Pyŏnhwadoen Yŏsŏng-ŭi Salm" [Women's Changing Lives under Japanese Colonialism]. In *Uri Yŏsŏng-ŭi Yŏksa* [A History of Korean Women], by Han'guk Yŏsŏng Yŏnguso [Institute for Research on Korean Women], 302–29. Seoul: Ch'ŏngnyŏnsa, 1999.

Smith, Sheila. *The Other Nation: The Poor in English Novels of the 1840s and 1850s.* Oxford: Oxford University Press, 1980.

Smith, W. Donald. "The 1932 Aso Coal Strike: Korean-Japanese Solidarity and Conflict." *Korean Studies* 20 (1996): 94–122.

So Hyŏng-sil. "Singminji Sidae Yŏsŏng Nodong Undong-e Kwanhan Yŏngu" [Research on the Women's Labor Movement in the Colonial Era]. Master's thesis, Ewha Women's University, 1990.

Sŏ Ŭn-ju. "Kang Kyŏng-ae: Kungp'ip Soke P'iŏnan Sahoejŭŭi Munhak' [Kang Kyŏng-ae: Socialist Literature Blooms in the Midst of Poverty]. *Yŏksa Pib'yŏng* [History and Criticism] (Autumn 1992): 296–300.

Soh, Chung-Hee. *The Chosen Women in Korean Politics.* New York: Praeger, 1991.

Sŏk Chŏng-nam. *Kongjang-ŭi Pulbitt* [Factory Lights]. Seoul: Ilwŏl Sogak, 1984.

Song, Jesook. *South Koreans in the Debt Crisis. The Creation of a Neoliberal Welfare Society.* Durham, NC: Duke University Press, 2009.

Song Hyo-sun. *Sŏulro Kanŭnkil* [The Road to Seoul]. Seoul: Hyŏngsŏngsa, 1982.

Song Youn-ok. "Japanese Colonial Rule and State-Managed Prostitution: Korea's Licensed Prostitutes." *Positions* 5, no. 1 (Spring 1997):171–217.

Sprinker, Michael. *History and Ideology in Proust.* London: Verso, 1998.

———. *Imaginary Relations: Aesthetics and Ideology in the Theory of Historical Materialism.* London: Verso, 1996.

Suh, Dae-Sook. *Documents of Korean Communism, 1918–1948.* Princeton, NJ: Princeton University Press, 1970.

Suh, Ji Young. "The Cityscape as Seen by Factory Girls in Colonial Korea." Paper presented at the Association of Korean Studies conference, Leiden, 2009.

Suleri, Sara. "Woman Skin Deep: Feminism and the Post-Colonial Condition" In *Colonial Discourse and Post-Colonial Theory,* ed. Patrick Williams and Laura Chrisman, 244–56. New York: Columbia University Press, 1994.

Tomalin, Claire. *Thomas Hardy: The Time-Torn Man.* London: Penguin, 2006.

Tongil pangjik pokjik t'ujaeng wiwonhoe. *Tongil Pangjik Nodongjohap Undongsa* [The History of the Tongil Textile Labor Union]. Seoul: Tolbekae, 1985.

Tsurumi, E. Patricia. *Factory Girls: Women in the Thread Mills of Meiji Japan.* Princeton, NJ: Princeton University Press, 1990.

Wells, Kenneth. "The Cultural Construction of Korean History." In *South Korea's Minjung Movement: The Culture and Politics of Dissidence,* ed. Kenneth Wells, 11–30. Honolulu: University of Hawai'i Press, 1995.

———. "The Price of Legitimacy: Women and the Kŭnuhoe Movement 1927–1931." In *Colonial Modernity in Korea,* ed. Gi-wook Shin and Michael Robinson, 191–220. Cambridge, MA: Harvard University Press, 1999.

Williams, Raymond. *The Country and the City.* London: Penguin, 1973.

———. *Culture and Society.* London: Penguin, 1985

———. *Marxism and Literature.* Oxford: Oxford University Press, 1977.

———. *Problems in Materialism and Culture.* London: Verso, 1983.

Woo, Jung-En. *Race to the Swift: State and Finance in Korean Industrialization.* New York: Columbia University Press, 1991.

Wood, Marcus. *Slavery, Empathy, and Pornography.* Oxford: Oxford University Press, 2003.

Woolf, Virginia. *A Room of One's Own.* Oxford: Oxford University Press, 1992.

Yang, Yoon Sun. "Nation in the Backyard: Yi Injik and the Rise of Korean New Fiction, 1906–1913." PhD dissertation, University of Chicago, 2009.

Yi Hŭi-j'un. "Kang Kyŏng-ae Sosŏl Yŏngu" [Research into the Novels of Kang Kyŏng-ae]. *Han'guk Ŏnŏ Munhak* [Korean Language and Literature] (2001): 371–94.

Yi Jŏng-ok. "Ilcheha Kongŏp Nodongeso-ŭi Minjŏk kwa Sŏng" [Gender and Nation in Industrial Labor in Japanese Colonialism]. PhD dissertation, Seoul National University, 1990.

Yi Ok-ji. *Han'guk Yŏsŏng Nodongja Undongsa 1* [A History of the Korean Women's Labor Movement, vol. 1]. Seoul: Hanul Academy, 2001.

Yi Puk-myŏng. "Yŏgong" [Factory Girl]. In *Singminji Sidae Nodong Sosŏlsŏn* [Collected Labor Literature from the Colonial Period], ed. Ha Chŏng-il, 215–30. Seoul: Minjok kwa Munhak, 1988.

Yi Sang-gyŏng. *Han'guk Kŭndae Yŏsŏng Munhaksaron* [Rethinking Modern Korean Women's Literature]. Seoul: Somyŏng Haksul, 2002.

———, ed. *Kang Kyŏng-ae Chŏnjip* [Kang Kyŏng-ae's Complete Works]. Seoul: Somyŏng Ch'ulp'an, 1999.

Yi Sang-rok. "Siminul Song Pokhaenghanun Minjukuka, Taehan Minguk" [South Korea: The State That Sexually Assaulted Its Citizens]. In *20 Saege Yosong Sakonsa* [Scandalous Women of the Twentieth Century], by the Research Collective in Women's History, 227–38. Seoul: Yosong Simunsa, 2004.

Yi Sŏng-hŭi. "Hyŏndae-ŭi Yŏsŏng Undong" [The Contemporary Women's Movement]. In *Uri Yŏsŏng-ŭi Yŏksa* [A History of Korean Women], by Han'guk Yŏsŏng Yŏnguso [Institute for Research on Korean Women], 397–421. Seoul: Ch'ŏngnyŏnsa, 1999.

Yi Sŏng-ryong. "Ŏnu Yŏgong-ŭi Hasoyŏn" [A Factory Girl's Complaint]. *Tonga Ilbo* November 3, 1929, p. 5.

Yi Sun-gŭ. "Chosŏn Sidae Yŏsŏng-ŭi Il-kwa Saenghwal" [Women's Work and Lives in the Chosŏn Dynasty]. In *Uri Yŏsŏng-ŭi Yŏksa* [A History of Korean Women], by Han'guk Yŏsŏng Yŏnguso [Institute for Research on Korean Women], 192–224. Seoul: Ch'ŏngnyŏnsa, 1999.

Yi Yae-suk. "Yŏsŏng, Kudŭl-ŭi Sarang kwa Kyŏlhon" [Women, Their Loves and Marriages]. In *Urinŭn Chinan Baengnyŏn Dongan Ŏt'ŏke Salassŭlkka? 2* [How Did We Live over the Last 100 Years?, vol.2], 208–11. Seoul: Yoksa Pip'yŏngsa, 1998.

Yi Yi-hwa, ed. *Han'guk Kŭnhyŏndaesa Sajŏn* [Dictionary of Modern and Contemporary Korean History]. Seoul: Karam Kihoek, 1990.

Yi Yŏng-mi. "Taejung Kayo Sŏk-ŭi Padawa Ch'ŏldo" [The Beach and the Railway in Popular Songs]. In *Urinŭn Chinan Baengnyŏn Dongan Ŏt'ŏke Salassŭlkka? 1* [How Did We Live over the Last 100 Years?, vol.1], 96–114. Seoul: Yŏksa Pip'yŏngsa, 1998.

Yŏksa Munje Yŏnguso [Historical Issues Research Center]. *K'ap'ŭ Munhak Undong Yŏngu* [A Study of the KAPF Literary Movement]. Seoul: Yŏksa Pip'yŏngsa, 1989.

Yoo, Theodore Jun. *The Politics of Gender in Colonial Korea: Education, Labor, and Health, 1910–1945*. Berkeley: University of California Press, 2008.

Yu Dong-wu. *Onŭ Dolmaengi-ŭi Oech'im* [The Cry of One Stone]. Seoul: Ch'ŏngnyŏnsa, 1984.

Yu Jin-o. "Yŏjikkong" [The Textile Factory Girl]. In *Singminji Sidae Nodong Sosŏlsŏn* [Collected Labor Literature from the Colonial Period], ed. Ha Chŏng il, 271–310. Seoul. Minjok kwa Munhak, 1988.

Yun Jŏng-mo. *Koppi* [The Halter]. Seoul: P'ulbitt, 1988.

Index

CPSIA information can be obtained at www.ICGtesting.com
Printed in the USA
BVOW07s2109020215

385870BV00001B/12/P